T0231810

TBM TUNNELLING IN JOINTED AND FAULTED ROCK

TBM Tunnelling in Jointed and Faulted Rock

NICK BARTON

A.A. BALKEMA / ROTTERDAM / BROOKFIELD / 2000

Back cover illustration: Cutter change under a dangerous working environment (Shen et al. 1999).

Authorization to photocopy items for internal or personal use, or the internal or personal use of specific clients, is granted by A.A.Balkema, Rotterdam, provided that the base fee of US$1.50 per copy, plus US$0.10 per page is paid directly to Copyright Clearance Center, 222 Rosewood Drive, Danvers, MA 01923, USA. For those organizations that have been granted a photocopy license by CCC, a separate system of payment has been arranged. The fee code for users of the Transactional Reporting Service is: 90 5809 341 7/00 US$1.50 + US$0.10.

Second print

Published by
A.A.Balkema, P.O.Box 1675, 3000 BR Rotterdam, Netherlands
Fax: +31.10.4135947; E-mail: balkema@balkema.nl; Internet site: http://www.balkema.nl

A.A.Balkema Publishers, Old Post Road, Brookfield, VT 05036-9704, USA
Fax: 802.276.3837; E-mail: info@ashgate.com

ISBN 90 5809 341 7

© 2000 A.A.Balkema, Rotterdam
Printed in the Netherlands

Contents

PART 3
LOGGING, TUNNEL SUPPORT, PROBING AND DESIGN VERIFICATION

Preface

During a long international involvement with drill-and-blast excavations – engagements that started with the development of the Q-system in the early seventies – the writer has also been drawn into smooth bored TBM tunnels – at first with some reluctance it must be admitted. How could one classify the jointing properly when there was almost no overbreak? Why did low quality rock masses – where there was too much overbreak – cause such problems for the machines?

The writer's engagements with TBM projects has generally focussed on conflicts between Owner and Contractor, sometimes Contractor and sub-Contractor. Inevitably, these conflicts have related to the geology and hydrogeology – and their characteristic lack of concern for the TBM and its operator's wellbeing, not to mention that of all other parties involved.

The two geotechnical difficulties referred to above are specific to TBM tunnels. Lack of overbreak can easily mask the true rock mass quality and is actually a source of risk. When tackling ground that is of obvious bad quality, it appeared to the writer that the canopy or shield were actually hindering efficient pre-treatment – an opinion there is much support for in the literature these days.

However, and this is a very large however, the fact that TBM might advance more than 150 m per day, 500 m per week, 2 km per month or even 15 km in a year guarantees the admiration – almost incredulity – of any tunnelling engineer, and forces our acceptance of some TBM limitations.

This book is an attempt to quantify and understand both the records and the limitations in poor ground, whether from hard, abrasive and massive rock, or from faulted regions with erosion of fines and chimney formation. The tunnelling profession needs to explain the records and the occasional failures, and all the variable ground in between, which has most impact on prognoses.

.

Acknowledgements

The opportunity to work professionally in widely varying TBM tunnelling conditions with contractors, owners and consultants has been an essential ingredient in the development of the Q_{TBM} method.

For many reasons, including these widely different tunnelling conditions, the writer would like to acknowledge EuroTunnel, UK Nirex and GeoEngineering in the UK, Statkraft in Norway and Kashmir, Kraftbyggarna in Sweden, NOCON in Norway and Italy and Fuji RIC in Japan., not forgetting nature herself, and all the devious ways in which geology, hydrogeology and tunnel depth interact.

Discussions with NGI colleagues Fredrik Løset and Eystein Grimstad and direct application of some of their richly varied tunnelling experience, is also sincerely acknowledged.

The committed and professional expertise of Pat Coughlin and Marcelo Medina Abrahão in production of the text and graphics will facilitate the trouble-free application of Q_{TBM} for which I thank them sincerely.

Finally, I dedicate this book to Eda Quadros, who was instrumental in creating both the necessary atmosphere and opportunity to concentrate on only one project at a time, after many years of varied but enjoyable professional work at NGI. Both Professor Lineu Ayres da Silva and Professor Carlos Maffei of the University of São Paulo, played an indirect but generous part in this project, including the loan of an office without telephone or e-mail, for which I am thankful.

Part 1. Basic interactions between the rock mass and the TBM

CHAPTER 1

Introduction

The frontispiece shows a photograph from the world's first bored tunnel. It is also a good example of the influence of jointing on overbreak. The exploratory Beaumont Tunnel was driven in chalk marl in 1881-1882. The classic wedge-shaped fall-out caused by three joint sets intersecting the tunnel wall was also a source of major difficulties when the Channel Tunnel was driven by much larger machines some 110 years later. The Beaumont TBM of 2.1 m diameter achieved weekly advance rates of between 30 and 60 m (Varley & Warren 1996), a highly respectable effort but later overshadowed by much higher performances and a best week of 426 m, by the time the 8.7 m diameter twin-tube Channel tunnels were completed in 1990-1991 (Warren et al. 1996).

The fact that penetration rate (PR) can reach extremes of 10 m/hr in certain TBM projects, while advance rates (AR) as low as 0.005 m/hr (or even zero) are experienced on occasion, suggests the need for predictive models and a good understanding of the important variables. Perhaps in no other rock engineering activity is the need for rock classification so important.

The fraction of total construction time (U) that the TBM can be utilised for boring, where:

$$AR = U \times PR \tag{1}$$

is strongly related to rock conditions, but also to many other factors. Nelson (1993) suggested that penetration rate (PR) and utilisation (U) could be separated and each related to a rock mass classification.

This book is an attempt to do exactly this, but with the difference that PR and U are partly related to the same basic classification system (the Q-system) with necessary 'modification' of RQD (it must be oriented in the tunnelling direction). A new stress-strength term is used to capture some extra features of joint anisotropy and its interaction with machine capacity (cutter thrust F).

A new term Q_{TBM} is formulated in stages, starting from the Q-value (since rock mass conditions are so important) and finishing with rock-machine and rock mass-machine interaction parameters. Q_{TBM} is designed to allow PR to be estimated, or Q_{TBM} can be back-calculated from actual performance, when tunnelling begins. Advance rate AR is estimated using Equation 1, but with the important difference that U is recast as a time-dependent and rock quality dependent variable.

3

CHAPTER 2

Some basic TBM designs

TBM designs have evolved in a very creative way particularly during the last quarter of the 20th century, in an effort to tackle ever-wider ranges of rock conditions with improved control of advance rates. Two basic types of machine are illustrated in Figures 1 and 2, a so-called open machine and a shielded machine. Several hundred meters of back-up equipment are associated with units of this size. The total investment and delivery time for new machines emphasises the need for reliable prognoses of rock conditions.

Useful summaries of technical aspects of TBM operations have been given by Nelson (1993) and Fawcett (1993). A helpful reference for the common technical terms and machine characteristics is reproduced in Figure 3.

Based on the premise that a good figure (especially a labelled figure) is 'worth a thousand words', we will allow Fawcett's three diagrams in Figure 3 to speak for themselves. However, the omission of working platforms for probe drilling, pre-injection, bolting (spiling and radial) and shotcreting – an omission made for reasons of clarity – should be noted. In this respect the three-dimensional drawing of an Atlas Copco machine given in Figure 1 is helpful, as it emphasises the advantages of the 'open' machine concept for necessary temporary support facilities, including shotcreting (not shown).

In later chapters, pre-treatment, temporary support and permanent support methods will be discussed, and methods for their selection will be highlighted. The mechanical engineering triumphs of these remarkable machines will not be discussed in this book, we will rather emphasise the rock mass (and hydrogeological) challenges faced by the machines, their makers, owners and operators, who require great ingenuity if they are to tackle the chosen sections of the subsurface successfully.

The writer is well aware that 'TBM designs' should occupy several thick books in preference to inadequate coverage in a chapter or page of a slim volume, as here. It is fortunately up to others who are suitably qualified to provide the profession with the fascinating story of TBM design developments during the last 100 years or more.

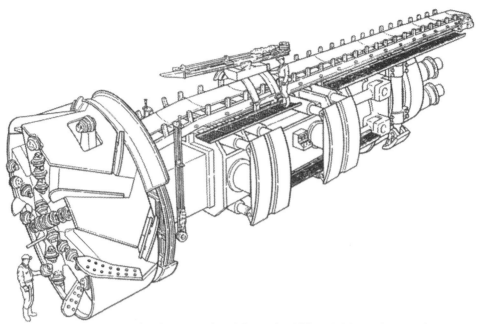

Figure 1. A 6.5 m TBM (Atlas Copco) equipped for probe drilling, bolting and overboring. (Nordmark & Franzer 1993).

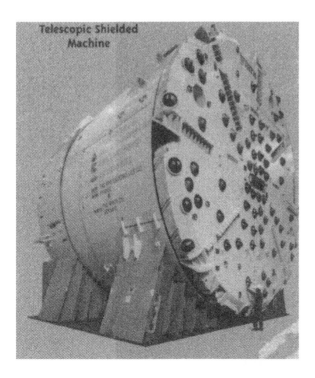

Figure 2. An 11.7 m TBM (Wirth Co.) with telescopic shield for faulted rock.

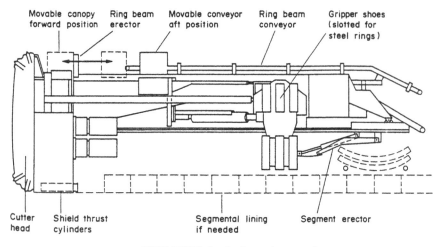

'NATM' TBM showing forward support features

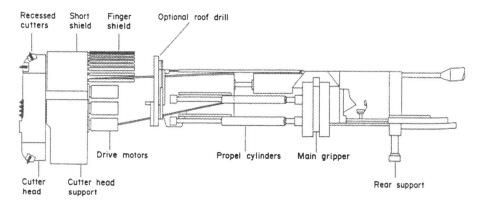

Typical rock TBM with short shield

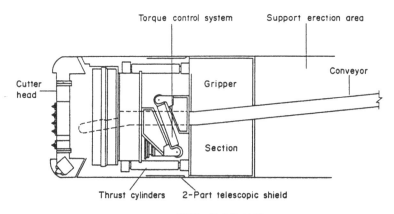

TBM with full shield

Figure 3. Three classes of TBM and common technical terms (Fawcett 1993).

CHAPTER 3

Summary of common geotechnical problems

TBM tunnels differ fundamentally from other types of tunnel (drill-and-blast, road header, hand-mined) due to the high level of machine-rock-rock stability interaction. In weak rock, especially if stresses are high, mucking and rock support may be the chief source of delay, and concrete element ring building may be compromised. In hard or very abrasive (quartz bearing) rock, equipment wear and cutter wear under the necessarily high loads will usually be the chief causes of reduced utilisation. Extreme water inflows may plague both on occasion.

The following list, two thirds of which is based on Nelson (1993), is a summary of the typical geotechnical and machine-rock interaction problems:
– installing support when there are wet and unstable rock conditions,
– cutter and cutterhead damage from face fallout, and cutterhead seizure,
– face fallout blocks are dragged around and increase overbreak,
– muck jams in the muck buckets and along the conveyors,
– removal of overbreak and eroded materials from the invert,
– unstable walls and overbreak that interact adversely with gripper function,
– rock or soil loads on the shield causing thrust losses, steering problems and delayed access to areas needing reinforcement or support,
– unstable invert causing steering problems and train derailment,
– erosion of fault debris, invert burial and flooding,
– chimney formation and block falls due to high inflows in faulted rock,
– damage to linings caused by fault void collapse,
– extreme cutter wear from abrasive rocks.
The fundamental geotechnical and machine-rock interaction problem is the delay caused by changed conditions and changed support needs. Efficiency drops on each occasion (Nelson 1993).

In highly jointed or faulted rock, the potentially high penetration rate may be greatly offset by the changed support needs, by rock jams, cutterhead seizure and subsequent gripper problems in the same zone. This is where a potential PR of 5 m/hr can become an advance rate of 0.5, 0.05 or even 0.005 m/hr as conditions become more and more extreme.

This brief summary of common geotechnical problems can serve as an introduction to the multitude of challenges that are faced in many of our longer or deeper TBM tunnels. In subsequent chapters some of the problems will be described in some detail. In particular, the challenges of fault zones, extreme water inflows and the difficulties with high stress and inadequately jointed rock are quantified and described with many examples.

CHAPTER 4

The TBM excavation disturbed zone

Some basic differences between TBM tunnels and drill-and-blasted tunnels are illustrated in Figure 4. These simple sketches from Wanner (1980) suggest a fundamentally different disturbed zone around such tunnels. In moderate to good rock conditions the zone with measurable change of properties (permeability, seismic velocity, deformation modulus) may be only some 10s of centimeters thick around a small TBM tunnel, but many times this value around a drill-and-blasted tunnel, especially under the invert. Overbreak of unstable rock also tends to occur during the blasting and does not need supporting to maintain the profile as it may in a TBM tunnel. Gripper problems that are unique to the TBM tunnel may be experienced in the zones prone to instability. These problems may occur only hours after excavation, or days or many weeks or even months later if the cutterhead gets blocked and major void formation occurs.

In the UK Channel tunnels, a wet, low-cover jointed region in the early kilometers caused many delays in ring-building due to the incomplete cut profile. As the concrete elements were not bolted, wedge block stability was compromised and a lot of backpacking was needed. Q-values (see Appendix) were typically in the range 5 to 10 (fair) in the 5 km of difficult ground, with Q-parameters frequently as follows:

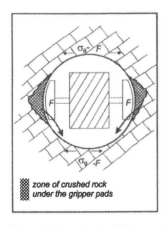
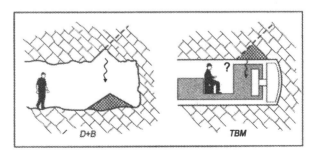

Figure 4. Some basic differences between TBM and drill-and-blast tunnels (after Wanner 1980).

$$Q \approx \frac{90}{6} \times \frac{1}{1} \times \frac{0.66}{1} = 10$$

Pre-construction sources of data, including the previously mentioned Beaumont tunnel, had indicated:

$$Q \approx \frac{100}{9} \times \frac{1}{1} \times \frac{1}{1} = 11$$

as the most frequently occurring quality (Barton & Warren 1995, Sharp et al. 1996). In each case, the non-dilatant joints and a sufficient number of joint sets indicated overbreak or need for early support. A shielded TBM may not provide this facility, and the shield, if too long or with advance too slow, will tend to carry this loose material. Delays with exposed ground cause a chain reaction, due to undesirable loosening of the newly exposed ground, which also tries to overbreak as support is delayed by ring-building problems.

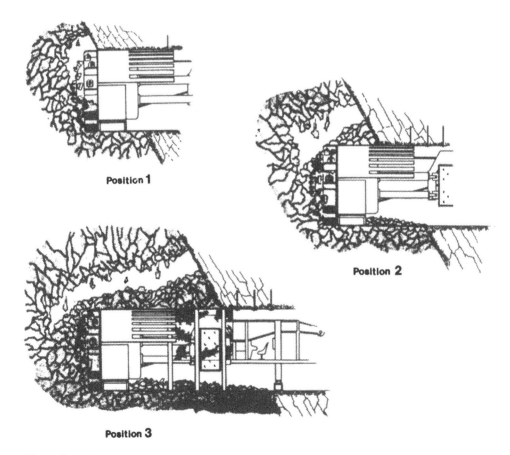

Position 1

Position 2

Position 3

Figure 5. Typical TBM problems in faulted rock with very short stand-up time (Robbins 1982).

At the Channel tunnel, progress averaged 127 m/week in this overbreaking ground and 338 m/week in the remaining 10 km of the UK excavations where Q-values were typically at least 40.

Figure 4 also illustrates the unique TBM problem of gripper load, which may damage the walls of the tunnel and cause roof instability in certain cases. In some projects, the modulus of the disturbed zone is actually measured by recording gripper deformations. However, the anisotropic tangential stresses around the tunnel profiles, which will be modified by the gripper forces, make interpretation less straightforward than desirable. A state of tension may be reached in the arch and invert, in cases where horizontal stress is less than the vertical stress, and the tangential stress σ_θ is inadequate. This may also increase the risk of block fallout. High stresses on the other hand make boring itself more time consuming due to the increased strength of the highly confined rock mass. This aspect will be discussed later.

A disturbed zone of more extreme character is illustrated in Figure 5, where penetration into unstable faulted ground has been taken too far without the necessary probe drilling and pre-treatment. Inadequate stand-up time allows material to cave (or to be eroded by water) at a faster rate than machine advance, and a void results. The void may take the form of a 'chimney' or 'inverted trench', or may extend the diameter of the tunnel on one or both sides. When a cylindrical 8 m diameter TBM tunnel becomes 12 m wide in unstable, sheared phyllites at 800 m depth, severe delays (e.g. 0.05 m/hr advance rates) may be difficult to avoid, especially if the TBM was actually designed for tackling hard quartzites. Q-values as low as 0.01 (or worse) may be involved and stand-up time proves inadequate for the length of shield and advance rate achievable.

CHAPTER 5

Basic factors affecting penetration rate

In this chapter, some of the basic factors such as rock strength and joint frequency will be examined, prior to more detailed treatment later. The conceptual probe drilling shown in Figure 6 penetrates a rock mass that has two levels of uniaxial strength (σ_{c1} and σ_{c2}) and two levels of joint frequency ($F1$ and $F2$). Experience shows that when drilling and when boring, the more massive and higher strength combination:

$$F_1, \sigma_{c2} \text{ and } Q_{1(max)}$$

gives the slowest penetration rates, while the fastest rates are given by

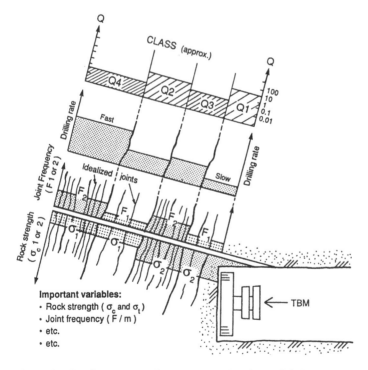

Figure 6. Joimt frequency, rock strength and Q-value and their conceptual relation to boring or drilling rate (Barton 1996).

F_2, σ_{c1} and $Q_{4(min)}$

Other combinations of σ_c and F (and Q) give intermediate rates.

Percussion drilling rates shown in Figure 7 (Thuro 1997) show a remarkable similarity in m/minute to the penetration rates of TBM in m/hr, when joint frequency increases as faulted rock is approached. This demonstrates rather clearly the potential benefits of probe drilling and measurement of data while drilling (MWD). The fact that a tendency for borehole collapse also precedes the fastest drilling rates has further parallels to pending TBM problems, when boring is going too fast.

According to Fawcett (1993), it is not unusual for a TBM to achieve three times the penetration rate in a softer rock compared to that which is achievable by the same TBM in harder rocks. However, as pointed out by the same author, the actual weekly progress may not vary so much as the different compression strengths of the various rock types penetrated. This is partly because of the greater needs for temporary support in the weaker rocks. The need for more frequent cutter change in the harder rocks and basically lower penetration rates are partially balanced by the generally more stable nature of the hard rock tunnel. A weak non-linear trend between PR and σ_c is indicated from the following mean data for more than 30 weeks of boring through shale, tillite and sandstone (after Fawcett 1993), see Table 1.

This simple but instructive set of data is plotted in Figure 8 in a format that will be used later. However, σ_c along the abscissa will be replaced by Q_{TBM}, which will include the ratio F/SIGMA, where F is the average cutter force in *tnf*, and

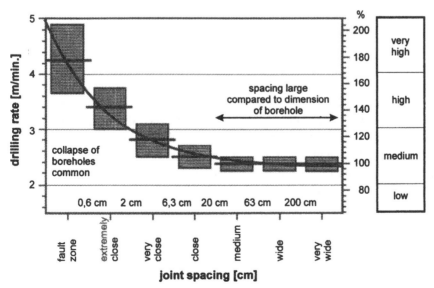

Figure 7. Percussion drilling rate and joint spacing in limestone show trends in (m/min.) that actually closely resemble TBM penetration in m/hr (after Thuro 1997).

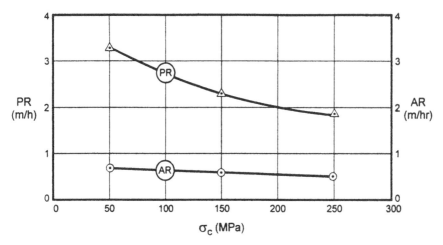

Figure 8. PR and AR data for shale, tillite and sandstone (derived from weekly average data given by Fawcett 1993).

Table 1. AR and PR data for rocks of different strength (derived from Fawcett 1993.

Rock type	σ_c MPa	PR m/hr	AR* m/hr
Shale	50	3.27	0.69
Tillite	150	2.32	0.57
Sandstone	250	1.88	0.50

*AR is based on weekly averages, converted to m/hr.

SIGMA is expressed in MPa. The term SIGMA for the rock mass will include σ_c, the uniaxial compressive strength for the rock material, the Q-value, the rock density and strength anisotropy.

Inevitably, uniaxial compressive strength is not the only parameter describing the rock matrix and its resistance to cutter penetration. Korbin (1998) refers to a granodiorite and a dolomite with almost the same uniaxial strengths (120-140 MPa) and tensile strengths, yet twice the PR in the dolomite. Abrasivity is clearly an important factor that needs to be reflected in prediction methods, even where PR is concerned, and quartz content and porosity may each play their role.

Brittleness and abrasion tests developed at the Technical University of Trondheim, and illustrated in Figure 9, are capable of distinguishing some of the mineralogical aspects that determine differences in penetration rate and advance rate, i.e. cutter wear. The following parameters are determined, as illustrated in Figures 9 through 12:

– Brittleness value (S_{20}),
– Siever's J-value (SJ),
– Abrasion value (AV and AVS),

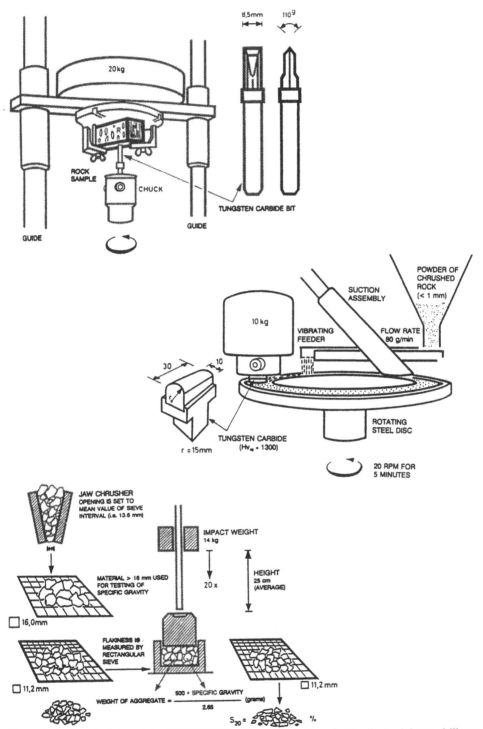

Figure 9. Basic principles of the NTH/NTNU brittleness test, Siever's (*J*-value) miniature drill test and abrasion test (Movinkel & Johannessen 1986, Bruland et al. 1995).

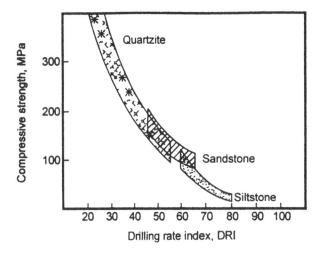

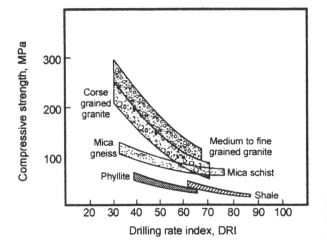

Figure 10. UCS and DRI trends for various rock types (Movinkel & Johannessen 1986).

– Drilling rate index (DRI),
– Bit wear index (BWI – used mostly for percussive drilling),
– Cutter life index (CLI),
– Joint and fissure class (Fig. 12).

The reader is referred to summary papers by Bruland et al. (1995) and Nielsen & Ozdemir (1993) for up-to-date details of these NTH/NTNU drillability tests, which have been widely used for hard rock tunnels. Bruland et al. (1995) have expressed the opinion that DRI and CLI (and also BWI) give a better distinction of drillability/borability rates between hard rocks than the uniaxial compression strength alone. However, their model assumes greater penetration rate with higher thrust per cutter, a relation that may be compromised in practice, because of the negative influences of high strength rocks if machine capacity is limited in relation to the 'toughness' of the rock mass.

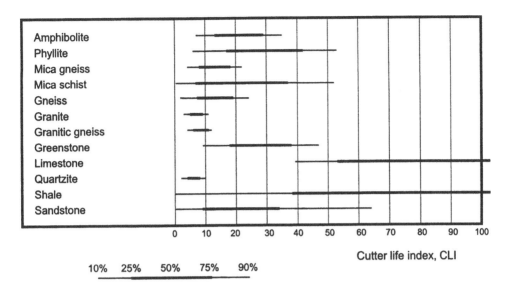

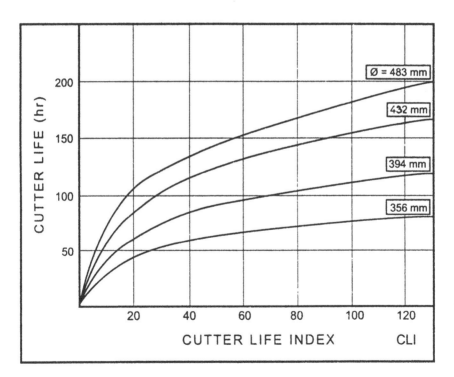

Figure 11. The cutter life index (CLI) and cutter life used in the NTH/NTNU TBM prediction model (Movinkel & Johannessen 1986, NTH 1994).

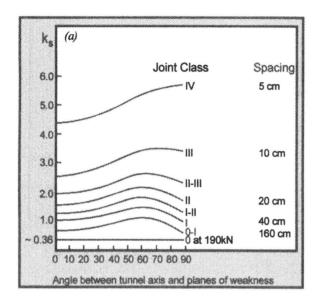

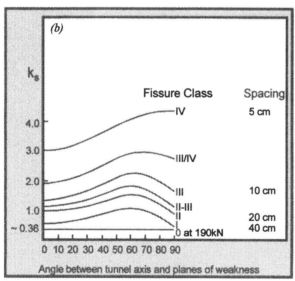

Figure 12. Correction factors for joint and fissure classes in the NTH/NTNU prediction model (Movinkel & Johannessen 1986).

CHAPTER 6

Penetration rate and thrust per cutter

An analysis of seven hard rock tunnelling projects in Norway reported by Barton (1996) considered the ratio of available thrust per cutter (F) and uniaxial compression strength (σ_c) as a potential measure of likely ranges of penetration rates. Since there was a range of rock compression strengths for each project, the ratio σ_c/F (MPa/tnf) also had a significant range. Net penetration rates also varied within each project.

The boxes and rectangles shown in Figure 13a represent the ranges for MPa/tnf and m/hr, while in Figure 13b the penetration rates have been normalised by TBM diameters, which ranged from 3.5 to 7.8 m. At least 120 km of TBM tunnels are represented.

The results appear to be logical, with high rock strength and low available thrust giving the lowest average penetration rates. However, many additional effects on penetration rate are ignored in such a simple analysis, including that of operator variability. Since typical Q-values were mostly in the range of 1 to 40, with no project showing especially high or low rock qualities, no further analysis of this data set is possible with the level of detail presented here.

The separate influence of uniaxial compression strength and thrust per cutter on penetration rates is shown in Figures 14 and 15. The first figure from Nelson et al. (1983) shows percentages of limestone at the face, while a TBM tunnel was progressing from a full face in shale ($\sigma_c = 68$ MPa) to a full face in limestone ($\sigma_c = 130$ MPa). Corresponding *reductions* in penetration rate are seen despite the increased thrusts per cutter (9 to 15 tnf) that were utilised as the stronger rock dominated the face. If the rock strength/thrust data is normalised as above (in the form σ_c/F MPa/tnf) the weaker shale shows a ratio σ_c/F of about 68/9 = 7.5 while the stronger limestone shows a ratio σ_c/F of about 130/14 = 9.3.

These two ratios would have suggested a penetration rate reducing from about 1.8 to 1.4 m/hr if the central trend in Figure 13 was used for prediction, while the actual result was 10.5 mm/rev reducing to about 7 mm/rev. The data set in Figure 15 showing penetration rate increases with thrust per cutter, for a given class of Hong Kong granite is also consistent with the above. A more or less quadratic relation between penetration rate and cutter force is suggested, with a rapid change of gradient beyond about 20 tnf.

However, when the available thrust per cutter is insufficient because of very

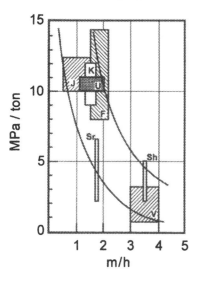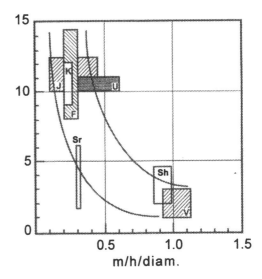

Figure 13. Average penetration rate data for 7 Norwegian TBM projects, as a function of σ_c/F (MPa/tnf) (Barton 1996).

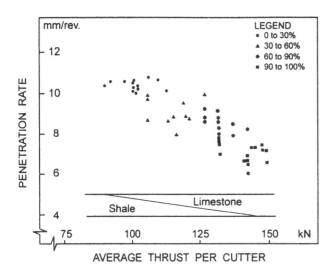

Figure 14. Penetration rate versus thrust per cutter for specified percentages of limestone (σ_c = 130 MPa). Remainder of face was shale (σ_c = 68 MPa) (Nelson et al. 1983).

strong rocks, the above logic appears to break down and penetration rate is seen to reduce with increased thrust per cutter, as shown for example by Grandori et al. (1995a) for Hong Kong granites (Fig. 16). However, when the penetration rate (i.e. mm/rev) is divided by the thrust (the more logical inverse of the field penetration index) a reducing trend with increasing uniaxial strength is seen (Fig. 17) which is logical and is consistent with the data trends shown in Figures 13, 14 and 15.

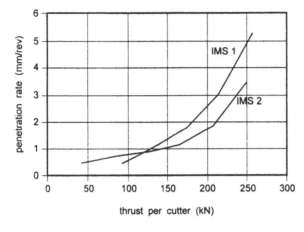

Figure 15. Penetration rate increase with thrust per cutter for individual classes of granite (Grandori et al. 1995a).

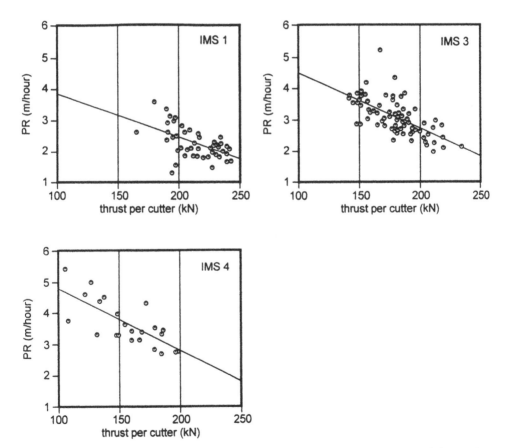

Figure 16. Penetration rates increase in poor rock classes, but reduce with increased thrust when some of the rock is hard to bore (Grandori et al. 1995a).

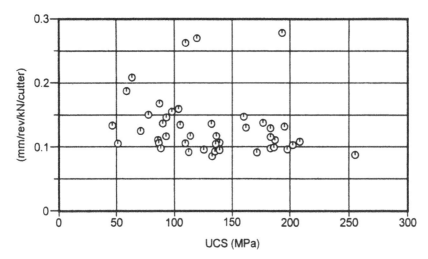

Figure 17. Penetration rate per thrust per cutter shows a general decline with increased uniaxial compression strength (Grandori et al. 1995a).

The TBM operator generally adjusts the thrust to the rock type. The actual PR may appear to fall despite higher thrust because the operator uses the highest levels of thrust when the rock is least borable. However, in a given rock a cutter thrust (F) of say 5 to 10 tnf might give only 1/7 to 1/10 of the PR achieved when the cutter thrust is increased to say 25 tnf. A quadratic relation between penetration rate and thrust (i.e. $3 \times F$ gives $9 \times$ PR) may be roughly assumed, and this simple 'rule-of-thumb' will be utilised later.

CHAPTER 7

The possible influence of stress and strength ratios

An interesting analysis of four hard rock tunnels in California was given by Klein et al. (1995). The σ_c or UCS range was 70 to 489 MPa, TBM diameters ranged from 3.4 to 4.3 m, and tunnel lengths ranged from about 2 to 5 km. The authors utilised the field penetration index (FPI) which is defined as follows:

$$FPI = \frac{thrust/cutter\,(kN)}{penetration\ rate\,(mm/rev)} \tag{2}$$

High values indicate greater difficulty of boring, usually due to high values of compression strength. However, the authors were not satisfied with UCS trends on their own, and therefore utilised the ratio of overburden stress (i.e. σ_v) and an assumed value of rock mass strength (UCS_{rm}) obtained from a Hoek & Brown (1980) formulation (see Fig. 18).

The term UCS_{rm} used by Klein et al. (1995) was derived from an early Hoek & Brown (1980) assumption

$$UCS_{rm} = \sqrt{s \times UCS^2}$$

For intact rock $s = 1$, while for jointed rock the authors used the following table from the same source, which is probably very conservative at the most jointed end of the spectrum of rock quality (see Table 2).

The above equation for UCS_{rm} will be evaluated from the range of s-values given below (in Table 2), for comparison with a preferred model for rock mass strength. This is a modified version of Singh (1993).

Table 2. Correlation of s-values with joint spacing (after Hoek & Brown 1980, as used by Klein et al. 1995).

Fracture frequency	Average joint spacing	s
1. Massive	> 3 m	0.1
2. Slightly fractured	1-3 m	0.004
3. Moderately fractured	0.3-1 m	0.0001
4. Highly fractured	30-500 mm	0.00001
5. Crushed	< 50 mm	0

$$\text{SIGMA} = 5\gamma Q_c^{1/3} \tag{3}$$

where γ = density, and $Q_c = Q \times (\sigma_c / 100)$.

The following evaluation of UCS_{rm} is made for a range of uniaxial compression strengths (see Table 3).

The higher values of UCS_{rm} listed in this table are presumably the dominant source of data in Figure 18, since the extremely low values of UCS_{rm} with increased jointing would predict gross overstressing with even a limited overburden.

Table 3. Estimates of rock mass uniaxial strength (Hoek & Brown 1980).

Fracture frequency class	s	UCS_{rm} values (MPa)			
		UCS =	30	100	300	MPa
1	0.1		9.5	31.6	94.9	
2	0.004		1.9	6.3	19.0	
3	0.0001		0.3	1.0	3.0	
4	0.00001		0.1	0.3	0.9	
5	0		0	0	0	

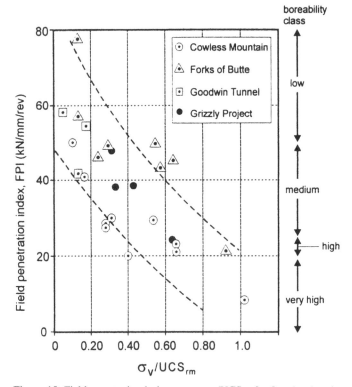

Figure 18. Field penetration index versus σ_v/UCS_{rm} for four hard rock tunnels (Klein et al. 1995).

The effect of confinement seems to be absent from the Hoek & Brown (1980) formulation used here, and therefore gives the very low values. However, the estimates of UCS_{rm} are perhaps consistent with a truly unconfined situation, which nevertheless does not usually correspond to conditions at the tunnel face, where a biaxial stress state is probably dominant, despite the localised TBM thrust through each cutter.

In spite of the above analysis of Klein et al. (1995), when rock is massive and highly stressed it will not be easier to bore, unless limited stress-fracturing occurs during the boring process. Tarkoy & Marconi (1991) describe an early case record from the Mont Cenis tunnel, where delayed 'popping' of the rock during pauses for re-gripping, made very difficult boring somewhat easier. However, in another case at the Star Mine in Idaho, stress-induced slabbing caused problems at the cutter head and penetration rates were only about 0.1 m/hr. In this case, the TBM was inadequate for the hard quartzite, and this method of drift development was abandoned.

It must therefore be assumed that the improved penetration seen in Figure 18 may be mostly a function of the favourable uniaxial strength and joint spacing effects, and not so much a function of the overburden pressure. Use of a classification method that accounts for the ratio of cutter force and uniaxial strength, and also includes joint spacing effects and stress effect separately, will presumably come closer to performance prediction.

CHAPTER 8

Basic mechanism of chip formation with roller cutters

In previous chapters, it has proved convenient to review penetration rates as a function of cutter thrust (which is a compressive force) and as a function of the uniaxial compression strength of the rock. Although crushing and extremely high stress levels are obviously occurring under the rolling tip of the disc cutter, the formation of radial cracks must presumably be tensile failure, and the formation of chips will require crack propagation and also tensile failure. Some shearing and mixed mode behaviour must also be involved.

A visualisation of the chip formation process is shown in Figure 19 from Rostami & Ozdemir (1993). There is an optimum cutter spacing for intact rock, which may not remain optimal when joints are involved as well. There is also an optimum joint spacing for a given cutter spacing.

An added complication is the speed of cutter rotation which obviously varies with its radial location in the cutter head. Smaller cutters rotate faster and larger machines will be rotated at lower revolutions/minute to limit otherwise exceptionally high cutter rotation speeds. Unfavourable effects on cutter bearings (and cutter wear characteristics) and less efficient breakage of the rock have been recorded when cutter rotation speeds are too high. Figure 11, showing cutter life in relation to cutter diameter (NTH 1994) is indirect evidence of the unfavourable effect of cutter rotation speed, but this figure also contains indirect evidence of the advantages of high cutter loads for which the larger diameters are essential.

When joints are present the cutting/chip/block formation is obviously significantly different, and more efficient, as implied by the joint-plane oriented 'overbreak' sketched by Aasen (1980) and reproduced in Figure 20. Key factors for faster penetration are the joint spacing, tensile strength and the joint or fabric orientation, each of which are treated separately later.

Since the tensile strength is usually linked to compressive strength, correlation between penetration rate and UCS may 'work' as we have seen, but could be improved. This is because rocks such as slate and schist, which may have unusually low tensile strengths, are exceptionally easy to bore in a particular range of orientations, but not outside this range. Tensile strength and orientation of fabric and jointing are the next parameters for this brief review.

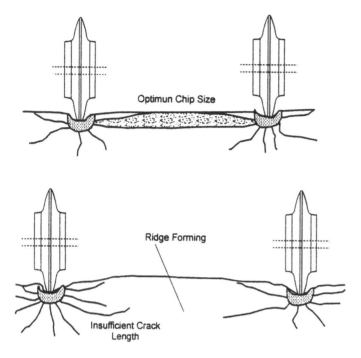

Figure 19. Chip formation in intact rock (based on Rostami & Ozdemir 1993).

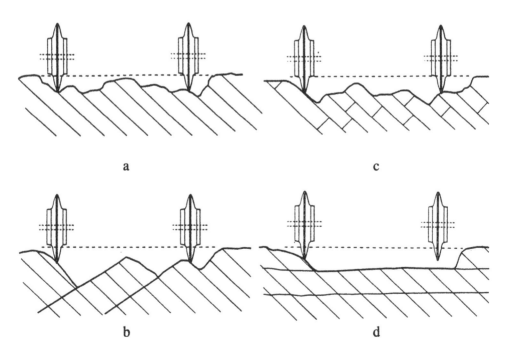

Figure 20. The favourable effects of fabric and jointing (based on Aasen 1980).

CHAPTER 9

Tensile strength and its anisotropy

The set of TBM performance data shown in Figure 21 from Klein et al. (1995) indicates the important role of the tensile strength (expressed as point load index I_{50}). Specifically, a low ratio of UCS/I_{50} gives low penetration rates (higher FPI), i.e. when the tensile strength is higher in relation to σ_c (or UCS) compared to the common ISRM conversion factor $\sigma_c/I_{50} \approx 24$ (for 50 mm diameter core).

This ratio really describes the 'toughness' of rock for boring (i.e. values less than 24), while materials like slate (values more than 24) would be described as easy to bore. In this connection, the values of σ_c/I_{50} given by Sanio (1985) emphasise the importance of orientation in relation to foliation, since the anisotropy of

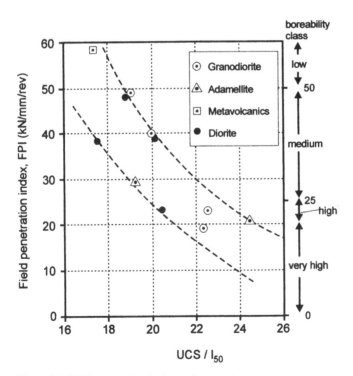

Figure 21. Field penetration index at four hard rock tunnels, and its relation to UCS/I_{50} (Klein et al. 1995).

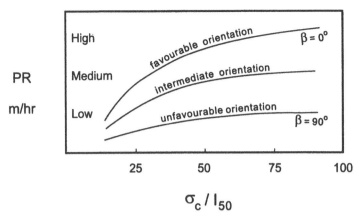

Figure 22. The expected general trends of PR with fabric orientation and the 'toughness' ratio σ_c/I_{50}.

Table 4. Ratios of σ_c/I_{50} for slate (from Sanio 1985).

		σ_c	I_{50}	σ_c/I_{50}
Slate	Parallel (‖)	104	1.3	80
Slate	Perpendicular (⊥)	189	8.4	22.5

I_{50} can be so marked (8.4 MPa compared to 1.3 MPa for instance), see Table 4.

According to Sanio (1985), six times faster penetration rates were recorded when a tunnel axis impingement angle β was 0° compared to 90° in these slates.

The trend for faster penetration rates with favourable ratios of σ_c/I_{50} may be approximately as shown in Figure 22. Note that a scale has not been given for PR (m/hr) as this will depend on many other factors.

There are in reality numerous parameters that impact penetration rate PR, and a wide range of simple graphs relating parameters that affect PR will be presented in the following chapters. These will include joint or fabric orientation, joint spacing and seismic velocity, which is related to several important rock mass characteristics. Strongest correlations will be seen when the multiple characteristics of rock masses in relation to TBM penetration have been captured in a comprehensive classification scheme.

CHAPTER 10

Penetration rate and fabric anisotropy

An extensive investigation of roller disc cutter penetration into anisotropic rocks reported by Sanio (1985) was designed to simulate as far as possible the cutter process illustrated in Figure 23. The author used 7 rock types [slate, gneiss, sandstone(s) and limestone] with varying degrees of anisotropy. He found that his experimental results correlated best with the anisotropic tensile strength of these rocks. The ratio of the point load index on 50 mm diameter core ($I_{50\perp}/I_{50\parallel}$) for perpendicular and parallel loading correlated directly with the variable cutter penetration achieved (P_0/P_{90}) where the subscripts 0° and 90° refer to the angle β in Figure 23.

$$\frac{P_0}{P_{90}} = \frac{I_{50\perp}}{I_{50\parallel}} \tag{4}$$

Sanio (1985) tested his results against other laboratory and in situ measurements with full-size cutters and obtained good correlations. Fourteen rock types with I_{50} values ranging from 1 to 22 MPa, and I_{50} anisotropy ratios ranging from 1 to 6.5 were included in the study.

When the anisotropic fabric (cleavage, schistocity, bedding) was at intermediate angles between $\beta = 0°$ and $90°$, a relatively smooth (flat S-shaped) transition was found, which actually resembles the anisotropic seismic velocity of such rocks. The maximum velocity is recorded parallel with fabric and the minimum perpendicular, e.g. Barton (2000), in much the same way as the ease of disc cutter penetration.

The experimental investigations of anisotropy effects by Sanio (1985) are given a realistic test and a convincing demonstration in a TBM tunnel through phyllonites, described by Aeberli & Wanner (1978). A 3.5 m diameter sewer tunnel passed through a syncline for about 140 m (Fig. 24) during which the angle between the tunnel axis and the foliation planes changed continuously from 60° through 0° and back to 60°. When the angle β was 90°, a PR of about 1 m/hr was recorded, and when 30°, PR had increased to 2.3 m/hr.

Based on Prandtl wedge concepts, Aeberli & Wanner (1978) suggested that the optimal angle β would be $45° + \phi/2$ where ϕ is the internal friction angle of the material. Shear failure of material as a passive wedge loaded from the sides of the cutter is perhaps a realistic mechanism in continuous plastic material (soft rock,

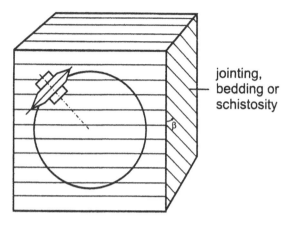

jointing,
bedding or
schistosity

Figure 23. Roller disc cutter penetration is least efficient when β = 90°, and depends strongly on the anisotropy of the point load strength indices (Sanio 1985).

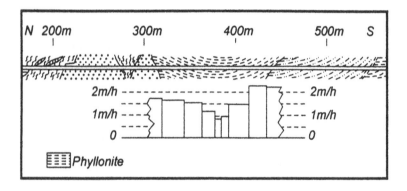

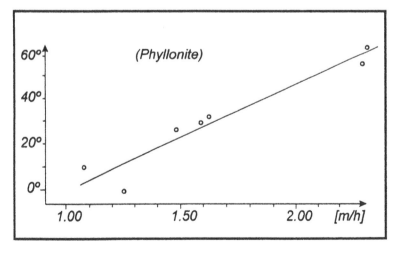

Figure 24. Graphic effects of foliation angle on penetration rates (average per day) in a 3.5 m diameter TBM tunnel in Switzerland (Aeberli & Wanner 1978).

clay zones, etc.) but is likely to be superseded by tensile crack formation and/or sliding on joint planes when more brittle and/or discontinuous rock is bored. A dilation angle will also be involved, making $45° + \phi/2$ a suspect model.

As pointed out by Nelson (1993), in the case of flat lying sedimentary rocks, a measure of joint spacing such as RQD may have little relevance to penetration rates, if the core is obtained from vertical boreholes. Efforts must clearly be made to interpret both the strike of principle joint sets and the spacing of principle joint sets in relation to the tunnelling direction. The same applies of course to aniso-tropic fabric such as schistocity or interbedded sediments with rapid cycling of properties.

An interpretation of RQD (and therefore the Q-value) with strict observance of the interpreted value parallel to the tunnel axis (we will call this RQD_o) will clearly be invaluable for better matching of the Q-value (or Q_o-value) to the PR-value. Ironically, a dominant structure parallel (or sub-parallel) to the tunnel axis will not only potentially reduce the value of PR, it may also cause a reduced utili-sation U, due to the need for longer sections requiring rock support and perhaps numerous problems with overbreak. In a TBM tunnel the overbreak may initially be carried by the shield or trailing fingers and can be exacerbated by the subse-quent gripper action if potential instability is not recognised and temporarily sup-ported (e.g. Fig. 4).

A dominant jointing or anisotropic structure parallel or sub-parallel to the tun-nel face will be of benefit both to PR, U and AR since this structure usually pres-ents something rather stable in a TBM excavation. Rock 'beams' in the periphery of a TBM tunnel represent stable structures, which in a drill-and-blast excavation would have tended to be damaged or 'over-broken' (appearing like a damaged dry-stone bridge arch). Overbreak in either type of tunnel may compromise sta-bility and will need support in both cases, as will be discussed later.

CHAPTER 11

Penetration rate, joint spacing and joint character

The extreme importance of jointing when σ_c is high can be judged by data given by Klein et al. (1995) for a section of TBM tunnel in granodiorite with $\sigma_c \approx 300\text{-}400$ MPa (see Table 5).

The frequency of jointing and its effect on TBM penetration rates has been documented on numerous occasions. Particularly graphic representations of the positive effects of joints is given in Figures 25a and 25b from Wanner & Aeberli (1979) and NTH (1994). The important effects of joint spacing are shown in Figure 26. However, Wanner & Aeberli (1979) noted that tight joints 'proved not to change the penetration rate, whereas open joints with gouge material lead to significantly higher specific penetrations'. A rough, tight or even healed joint [i.e. $J_r/J_a \approx (3{\rightarrow}4)/(1{\rightarrow}0.75)$ see Appendix] clearly gives a lot more resistance to block release in the tunnel face than one with the character and shear strength represented by for example $J_r/J_a = 1/6$ or $1/8$.

Wanner & Aeberli (1979) referred to penetration rates between 50 and 100% higher than the daily average when open joints with gouge were present. Shear zones with damaged wall rock were cited in particular as giving improvements in PR, but not necessarily in subsequent utilisation values due to the need for support.

Aeberli & Wanner (1978) indicated that rough-walled tension joints were least helpful to PR, especially if they were healed. On the other hand, shear features with associated damage were most productive of good penetration rates.

The basic influence of J_r/J_a in determining the degree of dilatancy (or its absence) is clear, and should be retained in any Q_{TBM} formulation. In principle $J_r/J_a > 1.0$ will reduce PR while $J_r/J_a < 1.0$ will tend to increase it. The opposite may be true for AR if U is adversely affected due to increased rock support needs. This will also depend on the number of joint sets and on the joint spacing which are captured quite well by the ratio RQD/J_n. The latter tends to relate to mode of failure, i.e. translational sliding (when blocks are larger) or rotation (when blocks are small).

As suggested earlier, an oriented RQD value, which we can write as RQD$_o$ must be specifically evaluated in the direction of tunnelling. Interpretation of vertical drill core will therefore require the experience of an engineering geologist, especially if oriented core is unavailable.

Table 5. An example of the effectiveness of jointing on improved penetration rate when hard rock is involved (Klein et al. 1995).

Degree of fracturing	mm/rev
Massive to slight	5
Slight to moderate	15
Highly fractured/crushed	30

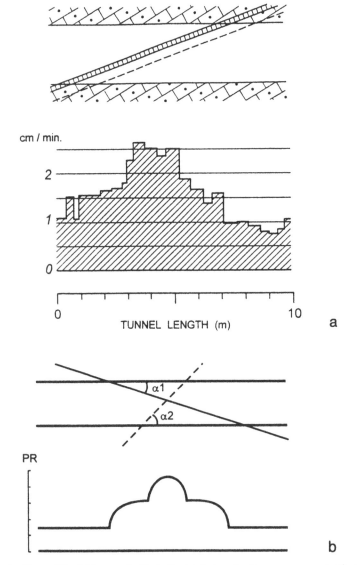

Figure 25. a) Measured effect of a jointed mylonite zone on penetration rate (cm/min) with constant cutter force (Aeberli & Wanner 1978), and b) Conceptual effects of two joints on measured penetration rate (NTH 1994).

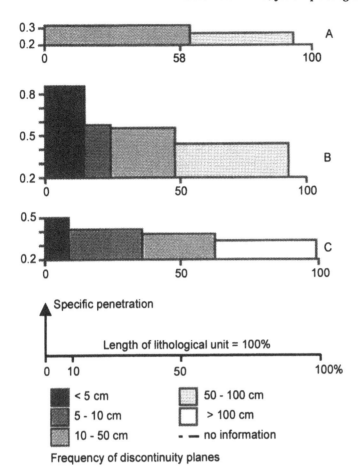

Figure 26. Effect of joint or discontinuity frequency on penetration rates through three limestones (after Aeberli & Wanner 1978).

When joint spacing approaches the cutter spacing, two to three times higher PR-values may be experienced. However, the orientation of joints or anisotropic fabric in relation to the tunnelling direction may have an even stronger influence. Sanio (1985) in fact showed up to six times higher PR-values with schistocity oriented parallel to the tunnel face than when oriented parallel to the tunnelling direction. The ratios of RQD_o/J_n, and J_r/J_a from the Q-system (see Appendix) are probably each operating here, together with a strong influence from the marked anisotropy of I_{50} in schistose rocks.

Part 2. Q, Q_{TBM} and rock mass variability

CHAPTER 12

TBM performance and rock mass classification

As late as 1993, Nelson was able to write that 'the geotechnical engineering profession clearly does not have a recommended method for quantitative estimation of the effects of rock mass variations on TBM penetration rate'. Yet rock mass characteristics generally influence overall tunnel boring machine performance more than the laboratory properties of the rock matrix because utilisation of the machine can be so severely affected in adverse conditions.

In this chapter, some classification methods that have been utilised in TBM performance analysis will be briefly reviewed. These should be evaluated in the light of what we have seen concerning the influence of:
– Uniaxial compressive strength (UCS),
– Point load strength (I_{50}),
– UCS/I_{50},
– Cutter loading (F),
– Orientation of fabric and joints,
– Anisotropy of I_{50},
– Joint spacing,
– Joint character (roughness, filling).

Grandori et al. (1995a) provided a very instructive analysis of a 7.4 km long, 3.6 m diameter TBM tunnel in Hong Kong. Using a rather simple joint spacing and weathering grade (IMS) classification, they showed consistent trends for PR, U and AR and average cutter thrust F in the predominantly fine and coarse granites, granodiorites and intrusive dykes. These analyses are shown in Figure 27. A comparative study for F and PR for a 5.4 km tunnel, also in granites is shown in Figure 28. In the latter, the PR value is seen to fall in the poorest rock class due to difficulties with the grippers. In the former case, only the advance rate falls in the lowest rock classes due to reduced utilisation and increased rock support needs. Note how the utilisation falls successively with lower qualities, likewise the necessary thrust for achieving generally faster and faster penetration rates.

In very approximate terms, the five IMS joint spacing – weathering classes can be compared to Q-values in the following manner:

IMS Class	1	2	3	4	5
Q-value	≥ 50	≈ 10	≈ 1	≈ 0.1	≤ 0.01

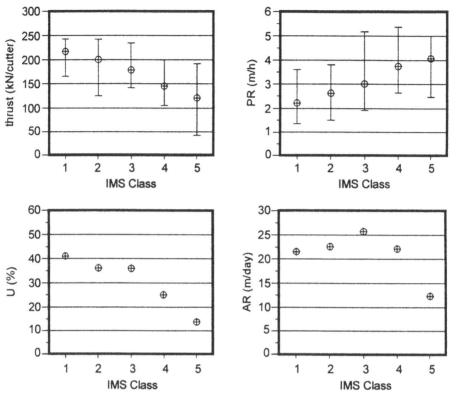

Figure 27. TBM performance data for a 7.4 km long tunnel in Hong Kong granites, granodiorites and intrusive dykes (Grandori et al. 1995a).

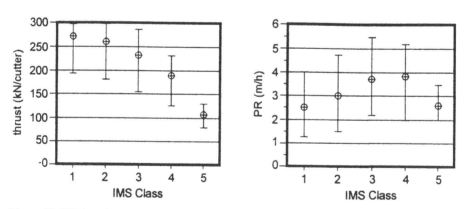

Figure 28. TBM performance data for a 5.4 km long tunnel in Hong Kong granites (Grandori et al. 1995a).

The potential application of the first four parameters in the Q-system ($RQD/J_n \times J_r/J_a$) was discussed by Innaurato et al. (1991). The Q-σ_c-PR correlations shown in Figure 29 were derived from measurements in 3.5 m diameter tunnels. The same authors converted their original RSR (= rock structure rating) observations in the dolomite and limestone (Wickham et al. 1974) to these approximate Q-values. The data suggests that PR correlation with Q-values would be improved if Q-values were normalised by uniaxial compression strength (i.e. $Q_c = Q \times \sigma_c/100$, Barton 1995), and probably further improved if cutter load in relation to σ_c was also introduced. These modifications are made later.

Another form of rock mass classification is the use of a P-wave velocity (V_p) obtained from shallow refraction seismic measurements performed along the tunnel wall. In the investigation phase, interpretation would need to be made from surface measurements, with depth, porosity and compression strength evaluation if a method such as the Q-value should be used for interpretation (see Barton 2000).

Mitani et al. (1987) showed quite a significant relationship between PR and V_p measured in two small diameter TBM tunnels. The data trend is shown in Figure 30. The authors also showed that V_p and rock class, and Schmidt hammer and rock class were clearly linked to one another, implying the value of both in classification. In other words, V_p correlates with rock mass conditions, and the Schmidt

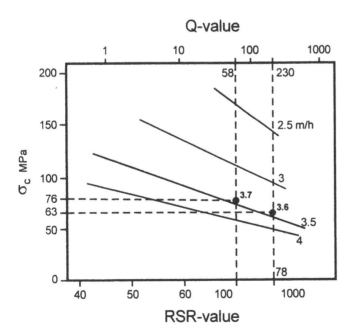

Figure 29. Correlation of PR with σ_c, RSR and an approximate Q-value, for a 3.5 m diameter tunnel (Innaurato et al. 1991).

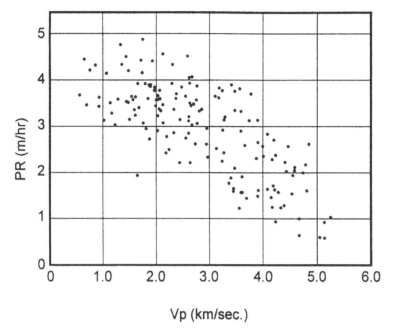

Figure 30. Correlation of PR and V_p at two small diameter TBM tunnels in Japan (Mitani et al. 1987).

hammer with rock strength conditions, and both correlated with PR and with the support needs.

A general equation PR = c-mV_p can be used to describe the PR-V_p behaviour shown in Figure 30. If we assume the general shallow seismic relationship for hard rocks: $V_p \approx 3.5 + \log Q$ (Barton 1991), then the following very approximate trend for PR (m/hr) can be derived from Figure 30:

$$PR \approx 2.7 - 0.8 \log Q \tag{5}$$

The rocks encountered at the two sites were predominantly granite, sandstone, slate and porphyry. Note that this simple empirical relationship gives average PR values in excess of 2.7 m/hr when the Q-value is less than 1.0, and smaller values when the Q-value is more than 1.0. When no joints are present and Q = 1000, an average PR value of 0.3 m/hr is predicted. This would be realistic if the rock was also hard and abrasive, as well as being massive. However, the cutter load in relation to the rocks bored is absent from such a simplified relation, and the simplicity should be regarded as a curiosity only.

In Chapter 5, two of the rock matrix parameters that are found to affect penetration rate (DRI and CLI, Figs 9 to 11) were described. Another part of the NTH/NTNU (Trondheim Technical University) method for predicting TBM performance that is shown in Figure 12, can be termed a 'rock mass classification method', or at least a partial classification method. It includes seven classes of

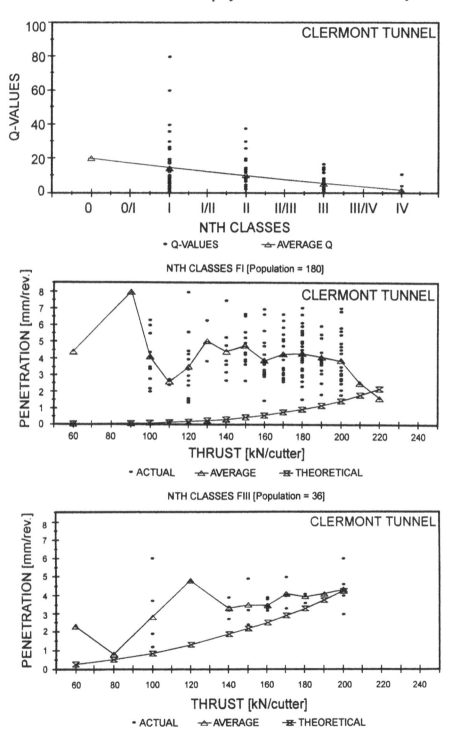

Figure 31. Evaluation of *Q*-value and NTH/NTNU joint and fissure classes and actual penetration rates at the Clermont tunnel in South Africa (McKelvey et al. 1996).

joint spacing and seven classes of fissure spacing. Fissures are defined as the non-continuous joints that cannot be followed around the whole profile. The spacing categories range from 50 mm to 1600 mm, and massive. Correction factors are applied for cutter diameter and cutter spacing.

The optimum joint orientation defined by the angle between the tunnel axis and planes of weakness is assumed by Bruland et al. (1995) to be about 60° for the joint spacings more than 100 mm, rising through 70° to 90° when the spacing is less than 100 mm and as low as 50 mm respectively. This is presumably a semi-empirical assumption. Joint character such as roughness, filling material or weathering are not specifically included, which may be a source of error. The number of joint sets are also excluded, so support related delays affecting utilisation will not be captured by the method when it is applied to poor rock conditions.

A detailed analysis of NTH/NTNU joint and fissure classes and comparison with RMR and Q-values was reported by McKelvey et al. (1996) during driving of two 5.5 km long aqueduct TBM tunnels in South Africa. Rocks encountered in the Clermont tunnel were very hard to extremely hard quartzitic sandstones, sandstones and siltstones, with an overall mean strength of 170 MPa. A comparison of RMR and Q-values and of NTH/NTNU joint and fissure classes is shown in Figure 31. Poor correlation is indicated between RMR and the joint/fissure classes and between Q and the joint/fissure classes.

Although an earlier generation NTH/NTNU prediction model was used by McKelvey et al. (1996), the scatter of actual penetration measurements compared to predicted emphasises the difficulties involved when many rock mass parameters are involved. The absence of the ratio UCS/F (MPa/tnf) is perhaps one of the reasons for the poor correlation, because penetration rate may actually decrease with increased thrust, as reviewed in Chapter 6. Unfortunately the authors did not attempt to use RMR or Q directly in correlations with penetration rate, but others have done so as shown in the next chapter.

CHAPTER 13

TBM performance and Q-system parameters

A unique set of TBM performance data and corresponding correlation levels with numerous standard rock mass description and classification methods was recently published by Sundaram et al. (1998). This is reproduced in its entirety in Tables 6 and 7. The TBM was instrumented, and an electronic data logger was installed to record standard machine variables. Five are listed in Table 6. The rock mass was continually mapped behind the back-up, with division into homogeneous structural and geological zones. Sixteen rock and rock mass parameters were measured, as listed in the table.

The *field penetration index* was determined from the ratio of thrust per cutter and penetration rate per cutterhead revolution. A total of 2825 m of TBM advance was analysed by these authors. The rock was medium to coarse-grained granite with σ_c = 130 to 246 MPa (mean 182 MPa).

The four strongest correlation coefficients for the *field penetration index* (which itself could be improved if it was inverted and divided by uniaxial compression strength – or even Schmidt rebound) were the volumetric joint count J_v (Palmström 1982) from which RQD was estimated, the ratio RQD/J_n, Q' = (RQD/J_n × J_r/J_a) and the Q-value itself. These correlations are shown in Figure 32 together with joint spacing (S) and Schmidt rebound (R). RQD derived from J_v showed good correlation with the field penetration index, but insufficient sensitivity below an RQD of about 75%.

The authors also separated the rock mass properties into zones where the thrust F per cutter was less than 12.5 tnf and where it was more than 12.5 tnf. This data is also reported in its entirety in Table 7. (The same numbered rock and rock mass properties apply as in Table 6). The field penetration indices had mean values of 25.7 and 64.5 kN/mm and utilisation was 22% and 37% in these two thrust per cutter categories.

Within the inevitable limits of a tunnel in one predominant rock type (granite), Sundaram et al. (1998) concluded that the average values of J_r/J_a in a selected zone gave better correlation with penetration index than the least favourable (shear strength and orientation) values usually used to estimate tunnel support needs (Barton et al. 1974). A separation is obviously needed here such that the conventional Q-value could be used to estimate utilisation (U) while a modified Q_{TBM}-value could be used to estimate penetration rate (PR).

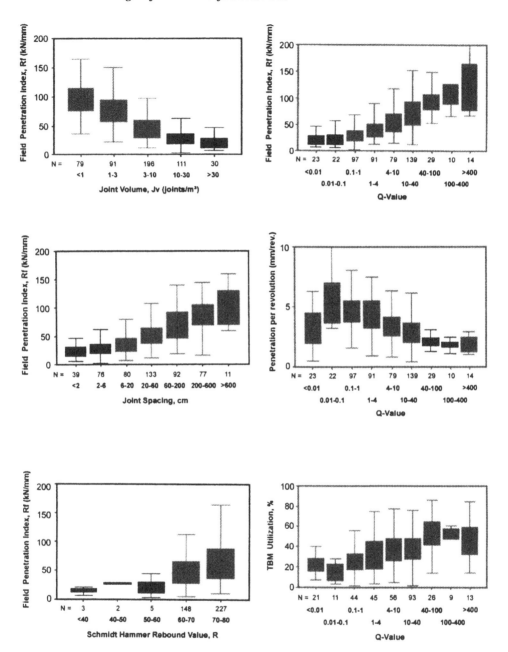

Figure 32. Box and whisker plots of TBM performance in granites, and its correlation to *Q*-value, *J*ᵥ, joint spacing and rebound *R* (Sundaram et al. 1998).

Table 6. Correlation values (r) of machine parameters with average rock mass properties.

Machine parameters	1	2	3	4	5	6	7	8
Field penetration index	0.33	0.53	0.22	0.17	-0.35	-0.19	0.72	0.61
Penetration per revolution	-0.29	-0.43	-0.16	-0.21	0.23	0.21	-0.54	-0.55
Specific energy	-0.21	0.20	0.15	-0.04*	-0.21	0.28	0.26	0.23
Torque	-0.09*	0.12	-0.01*	-0.09*	-0.19	0.15	0.11	-0.05*
Utilisation	0.15	0.29	0.25	0.09*	-0.20	0.11*	0.37	0.34

Machine parameters	9	10	11	12	13	14	15	16
Field penetration index	-0.76	-0.55	0.55	0.64	0.56	0.70	0.69	0.68
Penetration per revolution	0.63	0.46	-0.43	-0.52	-0.43	-0.57	-0.55	-0.54
Specific energy	-0.24	-0.24	0.19	0.27	0.20	0.23	0.24	0.30
Torque	-0.06*	-0.07*	0.15	0.10	0.17	0.11	0.14	0.12
Utilisation	-0.37	-0.34	0.25	0.32	0.24	0.38	0.39	0.46

*Insignificant correlation. 1 = Schmidt rebound value, R (L-type); 2 = Discontinuity alteration; 3 = Aperture (ISRM 1981); 4 = Infill material; 5 = Roughness (ISRM 1981); 6 = Joint orientation (Francis 1991); 7 = RQD (Palmström 1982); 8 = Fissure; 9 = Joint volume, J_v (Palmström 1982); 10 = Continuity; 11 = Rock mass alteration; 12 = Joint spacing (ISRM 1981); 13 = Strength estimate (ISRM 1981); 14 = RQD/Jn (Barton et al. 1974); 15 = (RQD/J_n) × (J_r/J_a) (Barton et al. 1974); 16 = Q-value (Barton et al. 1974).

Table 7. Comparison between average rock mass properties in areas below and above 125 kN thrust per cutter.

RMP*	≤ 125 kN per cutter	> 125 kN per cutter
1	66.53	71.22
2	Altered	Not altered
3	Narrow (0.6-2 cm)	Very narrow (0.2-0.6 cm)
4	Calcite, chlorite, fault gouge	Calcite
5	Smooth and undulating	Rough and planar
6	Favourable	Favourable
7	55	90
8	Fissured	Not fissured
9	21.27	5.55
10	50-100	0-50
11	Altered	Not altered
12	2-6 (very close)	20-60 (moderate)
13	Medium strong	Strong-very strong
14	6.95	38.12
15	2.48	54.39
16	0.861	47.680

*RMP = Rock Mass Properties – refer to footnote in Table 6 for explanation of properties.

In general terms, both the penetration rate (PR) and the advance rate (AR) show reduction with increased Q-value at the upper end of the Q range. However, as shown conceptually in Figure 33, there is a break-even quality when ease of boring is overtaken by poor stability and support needs.

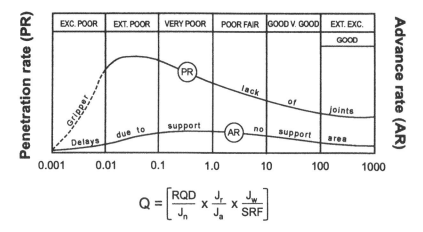

$$Q = \left[\frac{RQD}{J_n} \times \frac{J_r}{J_a} \times \frac{J_w}{SRF} \right]$$

Figure 33. A preliminary interpretation of PR and AR and the general influence of variable *Q*-values.

CHAPTER 14

TBM performance and the initial requirements for 'Q_{TBM}'

Several improvements must be made to the Q parameters incorporated in a 'Q_{TBM}-value' if it is to describe parameters of special relevance to TBM penetration rate. An oriented RQD-value, a more relevant J_r/J_a value and an SRF value that reflects cutter load in relation to rock strength are each quite basic requirements, and the list can be made longer.

The following basic model was tested first:

$$Q_{TBM} = \left(\frac{RQD_o}{J_n}\right) \times \left(\frac{J_r}{J_a}\right) \times \frac{J_w}{F/SIGMA} \tag{6}$$

where RQD_o is for an orientation parallel to the tunnelling direction, J_r/J_a are the relevant values for the discontinuities, joints or foliation planes that are most assisting the cutting process, F is average cutter load (tnf), and SIGMA is estimated as follows:

$$SIGMA = 5\gamma Q_c^{1/3} \quad \text{(modified from Singh 1993)} \tag{7}$$

It will be noted that the F/SIGMA ratio follows the spirit of SRF for competent rocks, in which the ratio σ_1/σ_c determines the SRF rating for rock support needs. For TBM penetration rates, the ratio σ_c/F (MPa/tnf) as shown in Figure 13 (Barton 1996) is obviously of importance. Inversion of this ratio and approximate conversion of σ_c to a rock mass strength SIGMA (with correction for density or porosity) would seem to be two important steps for better correlation with penetration rate.

The use of $Q^{1/3}$ by Singh in 1993 is interesting because of other relationships with $Q^{1/3}$ that have been recognised. In fact, the Q-value was developed by a form of back-analysis of rock mass strength and deformability as reflected in rock tunnel and cavern reinforcement and support needs. Excavation support pressures (P) obtained from case records show proportionality with $Q^{-1/3}$ (Barton et al. 1974), and the rock mass deformation modulus (M) shows proportionality with $Q_c^{1/3}$ (Barton 1995). In approximate terms when $M = 100$ GPa, $P \approx 1$ tnf/m^2, when $M = 10$ GPa, $P \approx 10$ tnf/m^2, and when $M = 1$ GPa, $P \approx 100$ tnf/m^2.

The six-order of magnitude scale of Q (0.001 to 1000) demands such a reduction in bandwidth (i.e. the cube root), which in the case of correlation with the seismic velocity of hard, near-surface rock masses was achieved by taking the logarithm of Q [$V_p \approx \log Q + 3.5$ expressed as km/s (Barton 1995)].

As will be demonstrated later, the fully developed term Q_{TBM} has an even larger bandwidth than the Q-value, and a suitable reduction factor must be developed by matching with case records.

Further progress in correlating TBM utilisation (U) to the Q-value, and penetration rate (PR) to a modified Q_{TBM}-value requires a large database. A first step in acquiring this database was a survey of penetration rates and advance rates (and utilisation from Equation 1) for well documented projects.

CHAPTER 15

The law of decelerating advance rates

In rock mechanics laboratories, small samples tend to show higher stiffness and strength than larger samples of the same rock. When tunnelling, the larger scale 'Weibull flaws' (now faults) become even more important. A short tunnel will have a certain distribution of rock qualities while a longer tunnel will in all likelihood add to both the centre and tails of the statistical distribution. Neither extreme ('exceptionally poor' nor 'exceptionally good') are desirable in a TBM project.

Not only the severity of local conditions but the length (i.e. width) of individual faulted zones can be expected (on a statistical basis) to increase as the tunnel length progresses from, say 1 to 10 km. The same can be said for the severity of potential water inflows. In effect the REV is increasing in size.

Since the TBM is already getting a beating and its rail system and muck conveyor belt is getting longer, there is also a mechanical-logistics statistic going hand-in-hand with the rock mass statistic. In this connection the statistic for some severe mixed face conditions and for both softer and harder rock as the tunnel length increases adds its own problems for vibration levels, cutter bearings, machine bearings and back-up wear-and-tear.

The net result is in fact a rather uniform logarithmic decay in advance rate and TBM utilisation where, in approximate terms, $T^{-0.2}$ governs the utilisation behaviour and T is expressed in hours. This trend will be seen shortly, from numerous projects (145 cases) involving more than 1000 km of TBM tunnels ranging from 2 to 12 m in diameter. A summary of the case record trends is shown in Figure 34, as an introduction to the decelerating advance rate concept.

Figure 34 shows that a typical penetration rate (PR) for an hour of uninterrupted boring of 3 m/hr becomes an average advance rate (AR) of 1.6 m/hr for 24 hours, then 1.1 m/hr for 1 week, 0.8 m/hr for 1 month and 0.5 m/hr for 1 year. This is quite normal TBM behaviour, and far different from the best day, week and month of 5, 4 and 3 m/hr (or even higher) achieved in record-breaking TBM projects. The reason for this is the successive decline in utilisation (U) where the following are the governing equations.

$$AR = PR \times U \tag{1}$$

or

$$AR = PR \times T^m \tag{8}$$

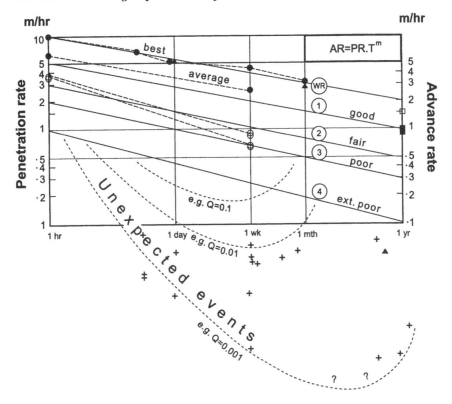

Figure 34. Declining average advance rate is seen as the unit of time (day, week, month) and tunnel length increase, based on 145 TBM tunnels totalling > 1000 km. The three black triangles refer to the Robbins (1982) case record for shale and glacial debris. Black circles refer to best and average results at Meraaker HEP (Johannessen & Askilsrud 1993). Squares refer to average and best UK channel tunnel results (Warren et al. 1996) at one year only. Crosses refer to diverse fault zones from widely different geologies. Open circles refer to a tunnel requiring systematic pre-injection (Garshol 1983).

where

$$U = T^m \qquad (9)$$

In the above example, the decay in utilisation shown in Table 8 is typically experienced. A 24-hour day is assumed since this is what governs the number of months or years that a project actually takes to complete. The fact that a working week may consist in practice of 100 or 120 hours is immaterial since 168 hours of activity or inactivity per week determines the actual tunnel progress. With a typical gradient (m) equal to –0.2, the approximate utilisation and advance rate is seen in Table 8.

Quotation of, for example, '36% utilisation' in a typical project description is meaningless, unless the time-scale is given, which is surprisingly infrequent in the TBM literature. In the above example, which is very typical, the time for completion of a tunnel project of 8760 m would be about two years, not the 47 weeks

Table 8. Typical PR, *U* and AR data for one year of TBM tunnelling.

	PR 1 hr	1 shift 8 hrs	1 day 24 hrs	1 week 168 hrs	1 month 720 hrs	3 months 2160 hrs	1 year 8760 hrs	
U =	100%	66%	53%	36%	27%	22%	16%	
AR=	3.0	2.0	1.6	1.1	0.8	0.6	0.5	m/hr

implied by 36% 'average utilisation'. On the other hand, if the mean PR could have been raised to 5 m/hr, instead of 3 m/hr, Figure 34 shows that 1 year completion would have been possible.

The lines marked 1, 2 and 3 in Figure 34 have been given the descriptions 'good', 'fair' and 'poor', and these adjectives reflect the 'coupled performance' of the TBM and the rock mass, and such factors as level of pre-investigation, machine design, support design and tunnel management.

These three lines were derived by Barton (1996) from a limited literature search of about 15 Norwegian and North American hard to medium hard rock tunnels in 1991. A much wider ranging literature search performed in 1999 including more than 1000 km of tunnels, did not give cause to modify the gradients or intercepts of these lines. The rather comprehensive source of TBM performance data shown in Figure 36 lies behind the trends shown in Figure 34. It will be noted that the large data set has been split into three parts:

a) Best performances (and various world records),

b) Average performances,

c) Bad ground performances (and unexpected events).

One type of 'bad ground' performance which is actually 'planned' is the delayed advance rate caused by the need to systematically pre-inject the ground to prevent or limit inflows, and thereby limit pore pressure drawdown and differential settlement damage caused by the presence of nearby clays. In Figure 34, the two open circles show mean PR values of 3.6 and 3.7 m/hr for two stretches of a 3.5 m diameter sewage tunnel near Oslo (Holmen right and left, Garshol 1983). The four circles and rectangular area shown on the '1 week' ordinate were due to the delayed advance rate (usually 50 to 63 m per 75 hour week) caused by the need for probe drilling and for 24 m long fans of holes for pre-injection of cement grout (when > 2.0 Lugeon water intake) and chemical grout (when the water intake was 0.2 to 2 Lugeon). When pre-grouting was not required, AR values about twice as high were achieved, i.e. 100 to 120 m/week. The delays caused by systematic pre-grouting are graphically illustrated in Figure 35.

Another form of 'planned' slower performance is the excavation of steeply inclined tunnels. Replacement of steep surface penstocks by TBM tunnelling at two hydroelectric projects in Italy, described by Astolfi et al. (1999), indicate more rapid deceleration ($-m$) for inclines than would be the case for horizontal excavations in the same rock. They recorded mean PR of 2.5 and 2.6 m/hr for the two projects, but only 0.55 to 0.60 m/hr for average daily advance rates.

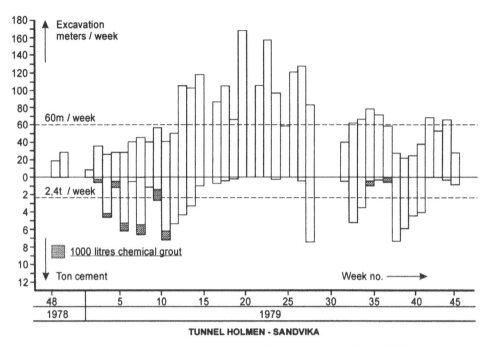

Figure 35. Effect of systematic pre-grouting on utilisation at part of the Oslo VEAS sewage tunnel of 3.5 m diameter (Garshol 1983).

Other data points plotted in Figure 34 include the world record (black circles) for best PR, shift, 24-hours, 1 week, and 1 month at a 10 km long tunnel for Meraaker Hydro Power Project in Norway (Johannessen & Askilsrud 1993). An average cutter load as high as 312 kN though 19 inch cutters was achieved with the high powered 3.5 m diameter Robbins HP TBM, driving mainly in phyllite (DRI = 60, Fig. 10), greywacke, greenstone, sandstone and some very hard meta-gabbro (DRI = 32). Only 140 rock bolts and 44 m³ of shotcrete were used for this 10 km long tunnel.

The black squares shown in Figure 34 are the average UK Channel Tunnel results for the 8.4 m diameter Robbins-Markham machines. The best day, week or month for these machines lie closer to the WR line. As already discussed, the early kilometers in more jointed and water-bearing chalk marl caused delays and average advance rates about one third of those achieved later. Strictly speaking it is perhaps surprising that soft rock machines (using shear picks rather than cutters) follow the trends of rock TBM with roller cutters. Mixed-mode 'fracturing' in both cases is perhaps the reason, plus the common deceleration trends of all TBM.

Other cases plotted in Figure 34 will be discussed in later chapters, particularly regarding 'unexpected events' and their dramatic temporary effects on advance rates.

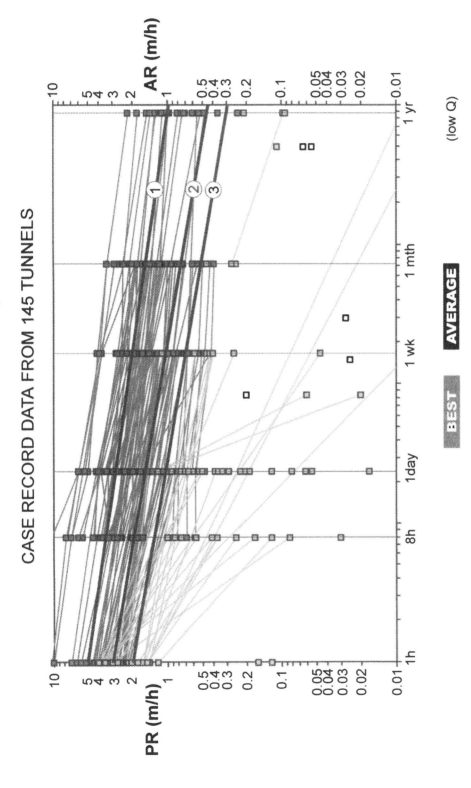

Figure 36. Eighty references, one including 27 tunnels, were the source of this best, average and bad ground TBM performance data from 145 TBM tunnels, totalling more than 1000 km.

CHAPTER 16

Utilisation and its decay with time

The three numbered lines 1, 2 and 3 in Figure 34 have gradients (m values in Equations 8 and 9) of –0.17, –0.19 and –0.21 respectively. A typical value of say $m = -0.20$ can be contrasted to $m = -0.15$ for 'learning curve' cases, where machine modifications and increased efficiencies are achieved. The lowest line in Figure 33 ('extremely poor') has a gradient $m = 0.25$. The pre-injection delays discussed above showed a mean gradient $m = -0.45$, which causes a dramatic loss of utilisation if greater efficiencies cannot be achieved. Figure 36 showed the direct effect of pre-grouting on utilisation very clearly. So far as it is understood, this was one of the very first TBM projects where systematic pre-injection was required to prevent settlement damage to buildings founded on nearby clay-filled depressions in the rock surface.

Table 9 shows the consequences of typical gradients (m) on utilisation values. The resulting decline in advance rates with time can be minimised by better TBM design and support methods, and by better conveyor systems. 'Waiting for the train' is an avoidable cause of declining AR, but was listed on several 'pie diagrams' concerning TBM utilisation seen during the extensive literature search.

It is to be hoped that the table below will encourage those reporting TBM case records to take care to record the time frame when they quote '37% utilisation'.(For all the gradients shown, whether 'learning curve', 'mean', or 'extremely poor', the time period quoted (day, week or month) is fundamental to the utilisation actually achieved. It is misleading not to qualify 'U' with 'T'. In the future perhaps one could consider utilising equation 9 directly ($U = T^m$) and give gradient (m) in addition to a specific utilisation factor for the average week or month.

Table 9. Estimates of utilisation (%) as a function of various utilisation decay gradients.

Period	Hrs	$m = -0.2$ Typical mean	m = –0.15 Learning curve	$m = -0.25$ Extremely poor	$m = -0.45$ Pre-inject. example
PR	1	100	100	100	100
shift	10	63.1	70.8	56.2	35.5
day	24	53.1	62.1	45.2	23.9
week	168	35.9	46.4	27.8	10.0
month	720	26.8	37.3	19.3	5.2*
year	8760	16.3	25.6	10.3	1.7*

*Such low values that improved efficiencies must be achieved.

CHAPTER 17

Unexpected events and their Q-values

A case record given by Robbins (1982) although fairly extreme, illustrates the problems and delays caused by unexpected conditions. The 3.1 m diameter TBM started an 8113 m long tunnel in the USA by boring 362 m of shale in less than 1 month (AR ≈ 0.56 m/hr). The next 270 m of completely unexpected water-bearing glacial debris took 203 days (AR ≈ 0.055 m/hr).

The remaining 7480 m, which fortunately was the shale again (σ_c ≈ 30-40 MPa) was achieved in 124 days (AR ≈ 2.5 m/hr). During the course of this record breaking advance, a best day of 127.7 m was achieved (5.3 m/hr) and a best month of 2.1 km (2.9 m/hr).

These two records and the extreme delay in the glacial debris are each shown in Figure 34 (see black triangles). They emphasise the need to relate PR and gradient (*m*) to suitable rock and rock mass classification schemes. They also emphasise the need for extensive pre-investigations (before the TBM is ordered) and the benefit of systematic probe drilling when pre-investigation is limited by high overburden. Deep mountain tunnels nearly always provide surprises. The case illustrated in Figure 5 from Robbins (1982) demonstrates the potential benefit that would be achieved by prior knowledge in the form of probe drilling.

When high overburden complicates prediction of tunnelling conditions, and there is perhaps insufficient reliance on seismic measurements because of anomalies, it is wise to expect occasional major problems. Folded flysch, limestones, chaotic sandstones and flysch and overthrust zones were each experienced at the Evinos-Mornos water supply tunnel in Greece where overburden rose to 1.3 km. This 4 m diameter tunnel of 29 km length was driven by four TBM, two of them open machines (similar to Fig. 3b) and two of them double shield machines with gripper shields (similar to Fig. 3c). According to Grandori et al. (1995b) some 34% of the tunnel was in RMR class V (RMR < 20) and 16% in 'RMR > class V' (RMR << 20). No squeezing rock term is found in RMR.

A graphic example of the tunnelling and stability problems encountered in a 'non-cemented mass formed by big sandstone blocks swimming in non-cohesive clay material' (plus high gas concentrations) is shown in Figure 37. Elsewhere the authors describe squeezing of more than 150 mm within 1 m from the face 'in almost zero time'.

In principle, the potential problems for a deep TBM tunnel resemble those for a

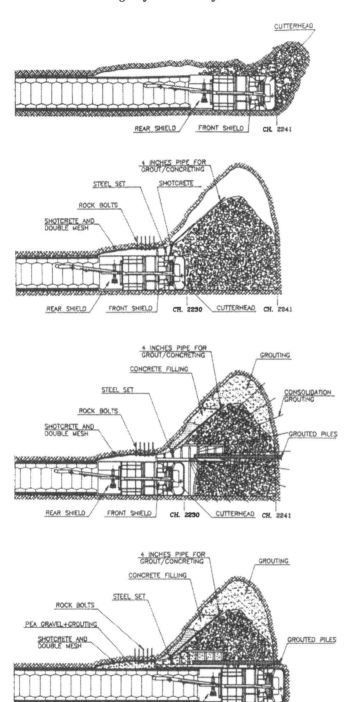

Figure 37. An extreme experience with a double shielded machine at the Evinos-Mornos TBM tunnel (Grandori et al. 1995b).

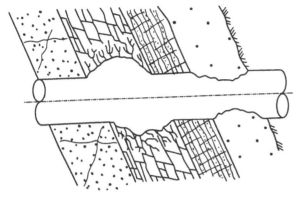

Figure 38. Inadequate stand-up time or squeezing problems (from the borehole stability literature) may create large scale problems for TBM (after Bradley 1978).

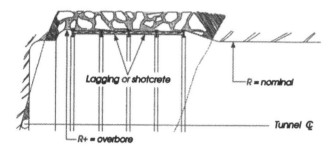

Figure 39. Overboring of a fault zone found by probe drilling to allow for subsequent heavier support (Toolanen et al. 1993).

deep borehole (Fig. 38). Potential overbreak or squeezing rock is supported by an appropriate mud weight in the one case, and by emergency measures (like measures shown in Fig. 37) in the other case. Prior knowledge of a zone with potential squeezing capabilities could, at least in theory, allow overboring to be performed as shown in Figure 39.

Grandori et al. (1995b) presented average daily advances for the four TBM which are shown in Figure 40. The rock mass class (RMC) was based on RMR and is given along the abscissa, with added classes 'RMC collapse' and 'RMC squeeze' as an extension of the RMR system. Some interpretations have been added to the original figure, including a tentative conversion to Q-value based on Barton (1995).

$$RMR \approx 15 \log Q + 50 \qquad (10)$$

The fact that 'still standings due to collapse and squeezing ground are not included' in the advance rates given in Figure 40 means that actual 'unexpected event' AR values less than 0.1 m/hr (as given in Figs 34 and 35) are removed.

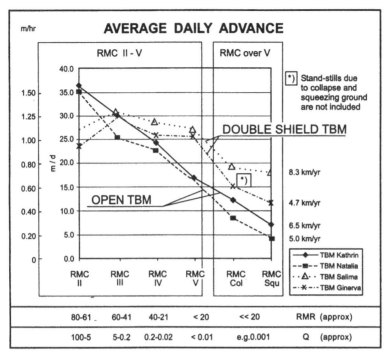

Figure 40. Average performance of four TBM at the Evinos-Mornos project (Grandori et al. 1995b), with additional interpretation by the writer.

The gradients of the m/day advance rates versus rock class are therefore less steep than they would otherwise be.

The actual performance of the open TBMs compared to the shielded TBMs is difficult to evaluate. However, Figure 40 leaves one in no doubt about the strong effect of rock mass quality on both types of machines.

Some of the most serious 'unexpected events' are naturally experienced where pre-investigations, for whatever good reasons, are limited. Logically this is frequently associated with deep tunnels or where there is inaccessible terrain. Careful analysis of air photographs and conversion of observations to approximate Q-values by experienced engineering geologists is a possible solution. However, air photographs may be unavailable for reasons of military security in some regions, and forests may be dense in tropical climates even over high ground. Localised mapping, geological interpretation and experience, are techniques that remain relevant under all conditions of inaccessibility.

In Taiwan, at the Pinglin tunnel, Tseng et al. (1998) recorded a total of 720 days of stuck machine time between 1993 and 1995 for the 4.8 m diameter pilot tunnel. The tenth delay since the TBM got stuck has lasted more than 1000 days by early 1999; this means five years of stoppage out of six years! The 11.7 m diameter main bores have also suffered massive problems, and drill-and-blast top heading advance is being used through two major fault zones (Wallis 1998).

A recent and graphic update of conditions at Pinglin, and some of the measures used by the contractor to try to control the exceptional conditions has been given by Shen et al. 1999. The back cover of this book also illustrates one of the typical situations encountered in this caving, abrasive, heavily jointed sandstone. The west-bound 11.7 m diameter TBM and back-up was crushed and buried over a distance of 120 m at the end of 1997, reportedly because of 45 m³/min inflow at almost 2 MPa water pressure.

At the Dul Hasti HEP in Kashmir, a section of the 8 m diameter TBM tunnel only 25 m in length took 270 days to traverse (AR = 0.004 m/hr). This delay had followed about 1 km of very tough boring through massive and very abrasive quartzites. An inclined bed of phyllite acting as a barrier was then traversed for some 50 m during which time a fault in the invert began seeping water (see Fig. 41). Deva et al. (1994) describe an unexpected blow-out of 4000 m³ of sands and pseudo-fluvial gravels from the invert nine days later, with an initial 70 m³/minute inflow of water and burial of the TBM and train. The source was a thick band of fractured quartzite extending to the mountain top. The *Q*-value in this problem zone has been estimated to have the following range, based on subsequent mapping:

$$Q \approx \frac{10-100}{6-15} \times \frac{1}{6} \times \frac{0.05}{5} \approx 0.01 - 0.03$$

This is the most extreme J_w-value (0.05) that is available in the *Q*-system, and it seems appropriate to the conditions described. In this special case, $J_w/\text{SRF} \approx 0.01$, which emphasises the danger of only using the first four parameters of the

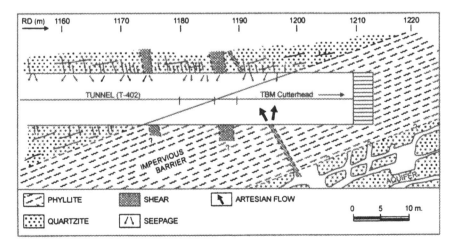

Figure 41. Blow-out location for 4000 m³ of sands and gravels and peak water inflows of 70 m³/min at Dul Hasti HEP, Kashmir (Deva et al. 1994).

Q-system, as done by some users of the Q-system, notably in Canada and Sweden.

Mixed face conditions, even with a double-shielded TBM, are a source of numerous potential problems, especially where hardness varies greatly. At the face, fall-out of hard blocks from a soft matrix (e.g. jointed basalt in clay as described by Hunter & Aust 1987) causes potential cutter head blockage, over-loading of the conveyor system and increased vibration. The instability created also affects support needs and subsequent gripper problems arise due to excessive displacements and fall-out of rock wedges. Thrust is limited, and auxiliary systems may be needed. The net result may be a 90% or more reduction in advance rate, even if TBM design is 'favourable' for such conditions.

In the Q-system, such mixed face conditions would usually be reflected by reduced RQD and J_r values, and by increased J_a and SRF values. A potential 99% reduction in Q-value (e.g. 5 to 0.05) would not be unreasonable, sometimes it could be even more.

CHAPTER 18

Water inflows in TBM driven tunnels

Water inflows cause various degrees of difficulty in TBM tunnels. Much depends on preparedness, and on whether or not discontinuity and fault infillings are washed out in the process. This may cause exaggerated overbreak and chimney formation, an unsafe working environment and require butt-to-butt setting of steel sets. Backfilling with concrete when peak flows subside adds to the potential delays. Faulted rock may not necessarily give major water inflows if a lot of clay is present, but adjacent jointing and fracturing may give very wet conditions.

At the record-breaking 13.5 km long Amlach Tunnel in Austria, Janzon & Buechi (1987) analysed TBM utilisation as a function of water inflow (l/s per 100 m). A 9 month long analysis shows *peak* monthly utilisation of about 65% when inflows were less than 40 l/s per 100 m. A steady reduction to peak monthly *U*-values of about 35% was experienced when inflows reached 100 l/s per 100 m.

These exceptionally high utilisation values at a record-breaking tunnel in dolomite (AR = 0.94 m/hr over 20 months) were achieved despite delays at two shear zones where high water inflows were experienced at the face. The two zones that were 4 m and 7.5 m wide took 150 hrs and 300 hrs to pass, due to the need for bulkheads and pumping of several hundreds of cubic meters of concrete. At a TBM site in Italy, a 2 m wide shear zone slowly crossing the tunnel for more than 50 m took 5000 hrs to pass, due mainly to frequent cutter-head blockage and massive outwash of fault-zone fines. An inverted fault trench and progressive chimney formation caused repeated burial of the invert and of the rails behind the backup.

At the UK Channel Tunnel, salt water inflows in the early, more jointed kilometers, caused TBM electrical operation problems and directly affected overbreak due to water pressure effects on the planar and least favourable joints and bedding planes. This caused a snowball effect, with difficulties in ring building and many unplanned delays as a result. Pre-injection of the running tunnels helped here.

At the Pinglin Tunnels in Taiwan, water inflows of up to 45 m³/min at a pressure of 18 bars were described by Tseng et al. (1998). At the Dul Hasti HEP in Kashmir, the water pressure encountered in 1992 was high enough to press the 8 m diameter TBM backwards by 40 cm, and peak inflow was more than 70 m³/min and remained above 10 m³/min for 165 days. Some 15-20 million m³ of water is estimated to have flowed from the zone depicted in Figure 41 during the first 5 years since this 'hydrogeological accident'. A new tunnel was needed to lead the water away.

It seems clear from several case records that a 'snow-ball effect' is experienced when weak materials occur alongside extreme water inflows. It seems to be appropriate that each of the six Q-parameters reflect such an accumulation of potential problems.

Besides unexpected flooding events as described in Chapter 17, the need to control water inflow with systematic pre-injection will usually be the chief source of constant delay (and increased gradient −m).

Improvements in drilling equipment, and its location on the TBM, are fundamental to minimising the necessary delays for pre-injection treatments, and improving its general level of quality. The need for post-injection is proof of a flawed system, which might also be caused by inadequate tunnel support.

CHAPTER 19

Consequences of limited stand-up time in TBM tunnels

Stand-up time inadequate for the chosen method of construction, and *inadequate strength in relation to stress* were the two top categories of geological problems identified in a US study of 87 underground projects in rock (Nelson 1993). If we add the third most frequent problem area – *unanticipated groundwater inflows*, it is clear that correct use of classification methods capable of representing these problems could help designers, contractors and owners in preparing for such problems. Improved pre-investigations are an unavoidable pre-requisite.

Conditions such as those illustrated in Figures 5 and 37 are not as rare as one might hope, and each clearly implies an extremely steep gradient (*m*) in Figure 34 which must be related to rock mass quality as implied in this figure. Insufficient stand-up time in the case of open machines (Fig. 5) and unexpected blocking of the cutter-head even in the case of double-shielded machines (Fig. 37) can lead to a snowballing situation, in which water and outwash of fines can be a major contributor. These conditions are not addressed by laboratory tests whatever their scale, and classification methods may also be pushed to their limits.

New machine concepts such as those illustrated by Robbins (1997) may go part of the way in tackling such problems, but probing and pre-treatment will be necessary when weak fault material and extremely high water pressures are both present.

For all three identified problem areas listed above from the US study, a combination of the RMR and Q-systems could have potential merit. The RMR addresses stand-up time and the Q-system addresses the ratio of stress to strength. Combining the two systems with a more workable conversion than that usually quoted in the literature is recommended as follows (see Equation 10, and Barton 1995):

Q	=	0.001	0.01	0.1	1	10	100	1000
RMR	≈	5	20	35	50	65	80	95

This conversion (which is only approximate due to the lack of squeezing, swelling and rock burst categories in the RMR method) has been added to the Bieniawski (1989) stand-up time diagram in Figure 42. Areas in this figure where the Q-value is less than 0.1 represent a critical region for TBM operation, because

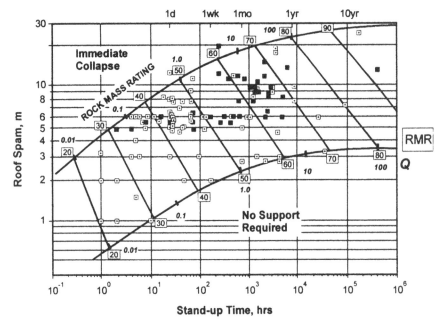

Figure 42. Bieniawski (1989) stand-up time estimates from case records. Conversion of RMR to *Q* from Equation 10 (Barton 1995).

shields may be too long in relation to driving rate, especially if previous bad ground is affecting gripper operations.

In the UK Channel Tunnel operations, several kilometers with *Q*-values in the range 2 to 6 (more than 100 sections) caused overbreak problems due to the considerable length of the shields. Sharp et al. (1996) discuss the adverse effects of between 15 and 18 m excavation-to-first support distances for the 5.4 and 8.4 m diameter machines used on the UK subsea drives, which also did not have the facility for building a bolted lining within the shields.

According to stand-up times in Figure 42, *Q*-values between 2 and 6 and roof spans (from the last support to the face) of 15 to 18 m as above, will be just inside the 'immediate collapse' zone, in other words while the TBM are driving these 15 to 18 m, possibly at reduced rates due to previous overbreak, new overbreak is gradually occurring which will be born by the tail end of the shields until it falls around the trailing fingers and causes perhaps a new cycle of ring building delays.

When *Q*-values are much lower, for example 0.01 or worse, conditions such as those illustrated in Figures 5 and 37 may cause almost immediate problems at the face and above and around the cutters. A local 50% increase in span (i.e. from 8 to 12 m) or a local 200% (or more) increase in height (i.e. from 5 to 15 m) can occur while one is boring or when the TBM is stopped due to cutter head blockage.

The stability may be compromised by the high erosive powers of high pressure water inflows, or by insufficient shear strength causing the tunnel wall to fail due

Figure 43. a) Sheared talcy phyllites at a collapse zone to the left of a TBM shield (NGI 1997).

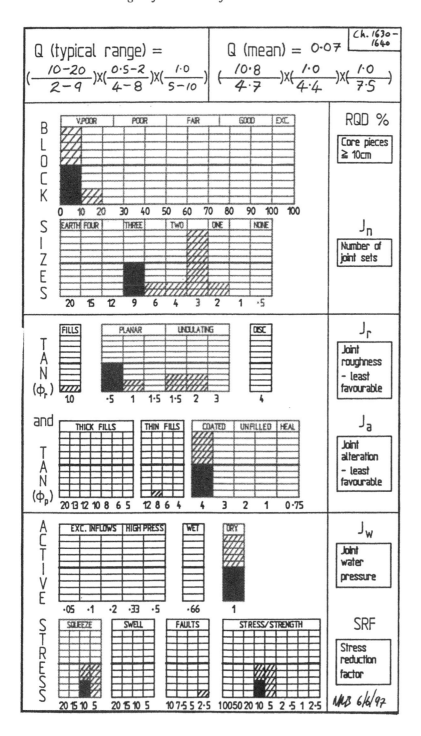

Figure 43. Continued. b) *Q*-logging of conditions shown in previous figure (black) (NGI 1997). (Traced from field logging sheets.)

Table 10. Approximate expectations of stand-up time according to Figure 42.

Q-value	No-support span (m)	Stand-up estimate
0.1	2	≈ 1 day
0.04	2	≈ 3 hrs
0.01	2	≈ 0.5 hr
0.1	5	≈ 4 hrs (6 m)
0.04	5	≈ zero (4.5 m)
0.01	5	≈ zero (3 m)

Numbers in parentheses refer to the no-support span when 'immediate collapse' is predicted.

to shearing or buckling of low friction layers (for example sheared phyllites or talc schists). This failure can occur despite (or because of) the delaying effect of the presence of the shield. The rock may be inaccessible for pre-support (spiling or pre-injection or shotcreting) unless some form of open machine is used (for example as shown in Fig. 3, types a and b).

Figure 42 suggests, in approximate terms, that Q-values in the range 0.1, 0.04 and 0.01 will have the stand-up times for 2 m and 5 m no-support spans that are shown in Table 10.

An example of Q-system logging of sheared phyllites with low stand-up time is shown in Figure 43 (see Appendix for description of Q-parameter ratings). A range of qualities from 0.007 to 1.0 is indicated, with a most typical value in a failing/collapsing zone of:

$$Q \approx \frac{10}{9} \times \frac{0.5}{4} \times \frac{1.0}{10} \approx 0.01$$

The prediction of 'immediate collapse' when the no-support span reached 3 m, after this amount of boring from the last flange-to-flange steel arches, appears to be a reasonably correct prediction of the conditions actually experienced, which are also illustrated in Figure 43.

CHAPTER 20

The relationship between PR, AR and Q_{TBM}

The development of a workable TBM performance prediction model is based on the following basic equations, which have been presented earlier:

1. $AR = PR \times T^m$ (see Fig. 34 and data trends in Fig. 35)

2. $Q_{TBM} \approx \dfrac{RQD_o}{J_n} \times \dfrac{J_r}{J_a} \times \dfrac{J_w}{(F/SIGMA)}$ (see Chapter 14, Equation 6)

3. $SIGMA = 5\gamma Q_c^{1/3}$ (see Chapter 14, Equation 7)

The terminology for the second and third equations is given in Chapter 14. The review of TBM performances in faulted rock gave convincing evidence of the need for a full Q-value, and SRF has therefore been 'reinstated' in the above quation 6 (see No. 2). The third term of Equation 6 can therefore be written as J_w/SRF (as usual), while the forth term is inverted for convenience and expressed as SIGMA/F.

The objective is to relate the penetration rate PR to the Q_{TBM} quality as defined above, but leave room for 'fine-tuning' of Q_{TBM} by additional factors such as σ_c/I_{50} (see Chapters 9 and 10, and Fig. 22).

The utility of $Q^{1/3}$ and $Q_c^{1/3}$ in tunnel engineering has been summarised earlier (Chapter 14). For the same reasons as those given earlier, $Q_{TBM}^{1/3}$ was investigated as a possible indicator of penetration rate PR (m/hr). On a log-log plot of PR versus $Q_{TBM}^{1/3}$, numerous sets of trial data gave a best fit gradient of approximately (–)0.66 and an intercept (at $Q_{TBM}^{1/3} = 1.0$) of 5. The equation therefore has the approximate form:

$$PR \approx 5Q_{TBM}^{-1/5} \qquad (11)$$

The component (–)0.2 was rounded from (–)2/9 = (–)0.222 for simplicity.

Since numerous data sets from more than 1000 km of TBM tunnels follow the 'law of decaying advance' ($AR = PR \times T^m$) shown in Figure 34, the following can therefore be derived:

$$AR \approx 5Q_{TBM}^{-1/5} \times T^m \qquad (12)$$

where T = total hours (24/day, 168/week, etc.).

It will have been noted from Chapter 16 (Table 9) that $m = (-)0.2$ is a typical mean value for TBM performance. Lines 1, 2 and 3 in Figure 33 have gradients of 0.17 (good), 0.19 (fair) and 0.21 (poor). A fair to good TBM performance may therefore be calculated from the following equation, which shows by chance equal power terms in relation to Q_{TBM} and time:

$$AR \approx 5Q_{TBM}^{-1/5} \times T^{-1/5} \tag{13}$$

where AR = expressed in m/hr.

Gradient (m) depends on the utilisation, which strongly depends on the coupled performance of the rock mass – TBM system, and how each affect each other. An excellent TBM design (e.g. high cutter loads, recessed cutters, closable buckets, efficient and high capacity conveyor, and appropriate temporary support close to the cutterhead) will tend to 'improve' the behaviour of an otherwise poor quality rock mass. Gradient (m) might be as good (i.e. low) as $(-)0.15$ to $(-)0.17$, and future machine designs will no doubt slightly reduce such gradients.

A poor TBM design will be accentuated and exposed by a poor quality rock mass and a high gradient (m) will result, for example $(-)0.22$ to $(-)0.25$. When the rock mass quality is abominable ($Q = 0.01$ to 0.001) due to squeezing pressures or high water inflows and 'zero' stand-up time, the poor qualities of the rock mass will inevitably bring down the complete system, however good the TBM design. A good example here is the Pinglin tunnel in Taiwan.

Temporary gradients (m) as large as $(-)0.7$ and $(-)0.9$ may even be experienced, implying very low (or zero) utilisation as the rock stability problem is tackled (for example Figure 37 from Grandori et al. 1995b). The 'unexpected events' shown in Figure 34 have gradients of this extreme magnitude, and will break the schedule and budget unless the coupled machine-rock mass problem can be solved within days or weeks or at worst a month. When problems are almost insurmountable (three such case records are shown in Figs 34 and 35), a new contract and a new contractor are sometimes the result, after six months or sometimes years of delay. Occasionally a new TBM is also used.

In Table 11, utilisation factors are suggested in relation to Q-values, based on the 'decaying or decelerating advance rate' gradients (m) discussed above.

On the basis of these gradients, Equations 11 and 12 have been evaluated, and the results are illustrated in Figure 44. For simplicity, it has been assumed that $Q_{TBM} \approx Q$, so that the gradients $(-)m$ from Table 11 can be used. More details are given in Table 12.

A suitable Q_{TBM} value for a hard, massive rock mass driven with an HP machine with high cutter loads might be the following:

$$Q_{TBM} = \frac{100}{2} \times \frac{2}{1} \times \frac{1}{1} \times \frac{85}{30} \approx 280 \quad \text{(tough boring)}$$

Table 11. Utilisation as a function of Q-values for average 1 month periods.

Q-value	0.001	0.01	0.1	1.0	10	100	1000
m	–0.9	–0.7	–0.5	–0.22	–0.17	–0.19	–0.21
U_{720}	0.003	0.01	0.04	0.24	0.33	0.29	0.25
	Unexpected events or expected bad ground. Many stability-related delays and potential gripper problems. Operator reduces PR			Most variation of (-)m is due to rock abrasiveness, i.e. cutter life index (CLI), quartz content and porosity are important. PR depends on Q_{TBM}			

Notes: 1) Monthly periods consisting of $24 \times 30 = 720$ hours are assumed (U_{24} or U_{168} would be much higher); 2) A 'learning curve' gradient might be as good as $m = (-)0.15$; 3) Systematic pre-injection stoppages may give a gradient as poor as $m = (-)0.4$ to $(-)0.5$ (e.g. $U_{720} \approx 0.07$ to 0.04 or only 4 to 7% boring time during each month).

Table 12. Estimates of PR and AR as a function of Q_{TBM} and assumed decaying advance rate gradients (m). Assume $Q \approx Q_{TBM}$, so that gradients $(-)m$ from Table 11 can be used.

Q_{TBM}	0.001	0.01	0.1	1.0	10	100	1000	
PR m/hr	(19.9)	(12.6)	7.9	5.0	3.2	2.0	1.2	m/hr
m	–0.9	–0.7	–0.5	–0.22	–0.17	–0.19	–0.21	assumed
AR m/hr	(0.05)	(0.13)	0.29	1.18	1.03	0.57	0.32	($T = 720$ hrs)
AR m/hr	(0.20)	(0.35)	0.61	1.62	1.34	0.76	0.40	($T = 168$ hrs)
AR m/hr	(1.14)	(1.35)	1.61	2.48	1.86	1.09	0.62	($T = 24$ hrs)

Note: Numbers in parentheses are theoretical as operator intervention is required here.

Notes:

1. $F = 30$ tnf (average cutter loads)
2. $\sigma_c = 250$ MPa and $\gamma = 2.7$ tnf/m^3
3. $Q = \dfrac{100}{2} \times \dfrac{2}{1} \times \dfrac{1}{1} = 100$, $Q_c = 250$
4. SIGMA ≈ 85 MPa (from Equation 7)
5. $RQD_o = RQD$ due to unfavourable joint angle, e.g. $\beta = 80°$ (Fig. 22)

A suitable Q_{TBM} value for a softer, well jointed rock requiring moderate cutter loads might be as follows:

$$Q_{TBM} = \frac{15}{15} \times \frac{1}{2} \times \frac{0.66}{1} \times \frac{6}{20} \approx 0.1$$

(favourable boring/unfavourable stability)

Notes:

1. $F = 20$ tnf
2. $\sigma_c = 25$ MPa and $\gamma = 2.1$ tnf/m^3

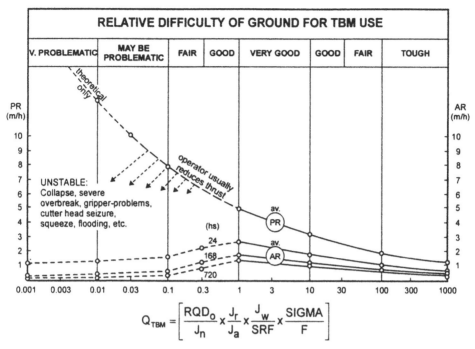

$$Q_{TBM} = \left[\frac{RQD_0}{J_n} \times \frac{J_r}{J_a} \times \frac{J_w}{SRF} \times \frac{SIGMA}{F} \right]$$

Figure 44. Log-linear versions of Equations 11 and 12 relating PR and AR with a preliminary expression for Q_{TBM}. (A final, fine-tuned version of this figure is given in the Appendix – Section A2.)

3. $Q = \frac{60}{15} \times \frac{1}{4} \times \frac{0.66}{1} = 0.66$, $Q_c = 0.16$

4. SIGMA ≈ 6 MPa (from Equation 7)

5. RQD_o = 15 due to favourable joint angle, e.g. β = 20° (Fig. 22)

The PR and AR ranges shown in Figure 44 are designed to match 'average' performances shown by case records in Figure 34. Obviously only one PR curve is shown (defined in m/hr with 1 hour as the unit of time) while three typical AR curves are shown. These are 'average' performances for 24 hrs (nominal 1 day), 168 hrs (nominal 1 week) and 720 hrs (nominal 1 month). The source of the curves (Equations 11 and 12) is given in Table 12.

The theoretical high penetration rate (and higher torque) in bad ground (i.e. Q or Q_{TBM} < 0.1) represents a danger to progress and is a source of uncertainty, since an operator would usually reduce the penetration rate dramatically when little resistance to penetration was registered with the required thrust.

Reference to Figure 35 reveals that *average* PR values are seldom above 5 m/hr, even though maximum rates of at least 10 m/hr can be achieved sometimes. Such a rate will usually exceed the capacity of the conveyor system and can be regarded as more theoretical than practical at present.

CHAPTER 21

Rock mass variability and its effect on predicted performance

Before adding 'complexity' (i.e. fine-tuning), the term Q_{TBM} and the gradient $(-)m$, an example application of this new TBM performance estimation will be given, in order to provide a 'feel' for the method. The following Q-value statistic for a 2 km long tunnel is assumed, based on pre-investigations (surface mapping, core logging, seismic refraction).

1. 1.0 km $Q_1 = \dfrac{100}{6} \times \dfrac{2}{1} \times \dfrac{0.66}{1} = 22$ ($\gamma = 2.7$ tnf/m^3)

          $\sigma_c = 150$ MPa $Q_{c1} = 33$

2. 0.5 km $Q_2 = \dfrac{50}{12} \times \dfrac{1}{2} \times \dfrac{0.5}{1} = 1.0$ ($\gamma = 2.5$ tnf/m^3)

          $\sigma_c = 100$ MPa $Q_{c2} = 1.0$

3. 0.4 km $Q_3 = \dfrac{25}{15} \times \dfrac{1}{4} \times \dfrac{0.66}{1} = 0.27$ ($\gamma = 2.4$ tnf/m^3)

          $\sigma_c = 75$ MPa $Q_{c3} = 0.2$

4. 0.1 km $Q_4 = \dfrac{10}{20} \times \dfrac{1}{6} \times \dfrac{0.5}{5} = 0.008$ ($\gamma = 2.1$ tnf/m^3)

          $\sigma_c = 10$ MPa $Q_{c4} = 0.0008$

From Equation 7, $SIGMA = 5\gamma Q_c^{1/3}$, the following estimates of SIGMA are made:

      $SIGMA_1 = 43.3$ MPa $F = 25$ tnf
      $SIGMA_2 = 12.5$ MPa $F = 20$ tnf
      $SIGMA_3 = 7.0$ MPa $F = 10$ tnf
      $SIGMA_4 = 1.0$ MPa $F = 0.5$ tnf

Cutter load (F) averages, which will be operator-influenced, are assumed values given next to each SIGMA value. If the simplifying assumption is made here that

RQD (conventional) = RQD$_o$ (oriented), the following values of Q_{TBM} can be estimated from Equation 6.

$$Q_{1\,TBM} = \frac{22}{25/43.3} \approx 38$$

$$Q_{2\,TBM} = \frac{1.0}{20/12.5} \approx 0.6$$

$$Q_{3\,TBM} = \frac{0.27}{10/7} \approx 0.2$$

$$Q_{4\,TBM} = \frac{0.008}{2.5/1} \approx 0.02$$

Following equation 11 for estimation of PR, the following are obtained:

PR_1 (1.0 km) $\approx$ 2.42 m/hr
PR_2 (0.5 km) $\approx$ 5.54 m/hr
PR_3 (0.4 km) $\approx$ (6.9 m/hr, theoretical)
PR_4 (0.1 km) $\approx$ (10.9 m/hr, theoretical)

Note: The two PR values given in parentheses would be reduced in practice by the operator if stability problems (such as in Fig. 5) were to be minimised. Delays due to pre-treatment needs caused by such conditions (very low Q-values) are estimated by the gradient $(-)m$, which increases towards a value of $(-)1.0$ when $Q = 0.001$, which would mean a need for alternative tunnelling methods, i.e. top heading and benching as at the Pinglin tunnels in Taiwan, if the Q-value could not be improved by pre-treatment.

The time T required to drive the tunnel a length L with a given advance rate is simply:

$$T = \frac{L}{AR} \tag{14}$$

Substituting Equation 8, we obtain:

$$T = \frac{L}{PR \times T^m} \tag{15}$$

from which the following can be derived:

$$T = \left(\frac{L}{PR}\right)^{\frac{1}{1+m}} \tag{16}$$

The 2 km long tunnel can therefore be expected to take the following length of time in the four classes of rock:

1. $Q_1 = 22$ (good)   1 km takes $(1000/2.42)^{1/0.82}$ = 1550 hours
   ($m \approx -0.18$ assumed from Table 12 or Figure 45) $\approx$ 65 days

2. $Q_2 = 1.0$ (poor/v. poor)   0.5 km takes $(500/5.54)^{1/0.78}$ = 321 hours
   ($m \approx -0.22$ assumed from Table 12 or Figure 45) $\approx$ 13 days

3. $Q_3 = 0.27$ (v. poor)   0.4 km takes $(400/6.9)^{1/0.64}$ = 569 hours
   ($m \approx -0.36$ assumed from Table 12 or Figure 45) $\approx$ 24 days

4. $Q_4 = 0.008$ (exc. poor)   0.1 km takes $(100/10.9)^{1/0.29}$ = 2086 hours
   ($m \approx -0.71$ assumed from Table 12 or Figure 45) $\approx$ 87 days

The above example demonstrates the sensitivity of the 'difficult ground' estimates to the value of PR (in this case an unrealistically high and hypothetical value) and to the value of $(-)m$ that is chosen as the 'decaying advance rate' gradient. A value approaching $m = (-)1.0$ for $Q \approx 0.001$ (worst possible rock mass quality) would bring progress to a complete halt according to Equation 8: $AR = PR \times T^m$ (Fig. 34) because when (m) approaches -1.0:

$$AR = \frac{PR}{T} \qquad (17)$$

This describes a drastic situation since the advance rate diminishes towards zero as T increases. For this reason a maximum negative gradient of $m = 0.9$ has been used in the foregoing. However, a situation such as that described by Equation 17 is not unrealistic of the extreme cases, since changed methods of excavation (non-TBM methods) are usually demanded by these extreme rock conditions (see for example Figure 37 where '$Q = 0.001$ conditions' have to be artificially improved before the TBM can resume progress).

At the Pinglin tunnels in Taiwan, well designed TBM machines have had to submit to 'conventional' top heading excavation and support, followed by slow TBM excavation of the remaining bench material. This is a project where the pilot tunnel has been bored for only one year out of six possible years (by 1999), and where continuation of the pilot bore is effectively blocked by the extreme conditions, and where start-up gets even less likely as more years go by (i.e. the Equation 17 'syndrome').

CHAPTER 22

Fine-tuning Q_{TBM} for anisotropy

Equation 6, which will now be reproduced in the modified form:

$$Q_{TBM} = \left(\frac{RQD_o}{J_n}\right) \times \left(\frac{J_r}{J_a}\right) \times \frac{J_w}{SRF} \times \frac{SIGMA}{F}$$

appears in 'normal ground' with normal cutter thrust (i.e. say 20 tnf) to give advance rates that agree well with the case records of average advance shown in Figure 35. The ratio F (cutter load)/SIGMA (rock mass strength) also appears to capture the important behavioural trait that penetration and advance rate may sometimes *reduce* as cutter load increases (due to the demands of cutting harder rocks). This important trend was shown in case records in Figures 13, 14 and 16, yet it does not appear to be incorporated in the NTH (1994) prediction model shown in Figure 31.

However, the F/SIGMA ratio does not yet capture some of the important details of behaviour. It is not yet sensitive enough to cutter force, because F is 'smoothed' by Equation 11, due to the negative quintuple root $(\ )^{-0.2}$. It also does not take account of the ratio σ_c/I_{50}. The importance of this ratio was emphasised by Sanio (1985) and Klein et al. (1995) (see Chapter 9 and Fig. 22).

There is a need to broaden the scope of the term SIGMA given by Equation 7.

$$SIGMA = 5\gamma Q_c^{1/3}$$

At present the rock mass quality Q has been normalised by $\sigma_c/100$ (i.e. when uni-axial compressive strength differs from the typical hard rock value of 100 MPa, either higher or lower Q_c values become more relevant). In a similar way it may be reasonable to normalise Q by the normal value of σ_c/I_{50} of about 24 for 50 mm samples. For convenience, this ratio will be rounded to 25, such that $Q_c = Q \times \sigma_c/100$ changes, by analogy to:

$$Q_t \approx Q \times \frac{25\,I_{50}}{100} = Q \times \frac{I_{50}}{4} \tag{18}$$

In other words, when the point load index differs from 4 MPa, higher or lower Q_t values will be used in a modified version of Equation 7.

For predominantly compressive failure processes we may utilise (as before):

$$\text{SIGMA}_{cm} = 5\gamma Q_c^{1/3}$$

where $Q_c = Q(\sigma_c / 100)$, while for predominantly tensile failure processes we may utilise:

$$\text{SIGMA}_{tm} = 5\gamma Q_t^{1/3} \tag{19}$$

where $Q_t = Q(I_{50}/4)$.

In a very weak talcy phyllite such as that illustrated in Figure 43 where a typical Q-value is about 0.01 (Chapter 19), we may have σ_c = 10 MPa (i.e. $Q_c \approx$ 0.001), while I_{50} may be as low as, say 0.04 MPa. In this case, $Q_t \approx 0.00001$. For equal values of γ, say 2.0 tnf/m^3, Equations 7 and 19 would give the following values:

Equation 7 $\sigma_{cm} \approx 5 \times 2.0 \times 0.001^{1/3}$ ≈ 1.0 MPa,

Equation 19 $\sigma_{tm} \approx 5 \times 2.0 \times 0.00001^{1/3}$ ≈ 0.2 MPa.

Considering an unfavourable joint inclination angle β (say 90°, see Figs 22 and 23), the dominant crushing failure caused by the TBM cutters might be dominated by SIGMA$_{cm}$, while a more favourable angle β (say 30°) would see tensile failure as the dominant mode, and the lower value of SIGMA$_{tm}$ would control behaviour. In the latter, a lower value of RQD$_o$ (oriented RQD) would also have been chosen, further accentuating the easier boring.

Of course the parameters SIGMA$_{cm}$ and SIGMA$_{tm}$ are only approximations to something that is extremely complex. They should nevertheless help to improve the predictive capabilities of 'Q_{TBM}'. As emphasised in Figure 22, when the joint angle β is favourable, the correct approach will be to use SIGMA$_{tm}$ since this emphasises the benefit of a potentially low value of I_{50} and a high ratio of σ_c/I_{50}. Conversely, when the joint angle β is unfavourable, the correct approach will be to use SIGMA$_{cm}$ since this emphasises the difficulty caused by high values of σ_c. 'Q_{TBM}' will therefore be written:

$$Q_{TBM} = \frac{RQD_o}{J_n} \times \frac{J_r}{J_a} \times \frac{J_w}{SRF} \times \frac{SIGMA}{F} \tag{20}$$

where

$\text{SIGMA} = \text{SIGMA}_{cm} = 5\gamma Q_c^{1/3}$ with unfavourable $\beta°$

$\text{SIGMA} = \text{SIGMA}_{tm} = 5\gamma Q_t^{1/3}$ with favourable $\beta°$

[where $Q_c = Q \times (Q_c/100)$ and   where $Q_t = Q \times (I_{50}/4)$]

The above preliminary 'fine-tuning' of z with consideration of I_{50} anisotropy, σ_c/I_{50} and $\beta°$ gives, together with RQD$_o$ the possibility of using a strongly anisotropic Q_{TBM} value, which may vary along the tunnel due to folding effects (see for example Fig. 24).

The following trends could be typical in a weak anisotropic rock with variable joint angle $\beta°$:

1. Unfavourable $\beta°$ is $> 60°$
    joint orientation RQD_o is higher   e.g. 60%
                                  SIGMA = $SIGMA_{cm}$ is higher   e.g. 1.0 MPa

2. Favourable $\beta°$ is $< 30°$
    joint orientation RQD_o is lower   e.g. 10%
                                  SIGMA = $SIGMA_{tm}$ is lower   e.g. 0.2 MPa

These particular examples of variations in RQD_o and SIGMA will cause Q_{TBM} (Equation 20) to vary by a factor of:

$$\frac{Q_{TBMU}}{Q_{TBMF}} = \frac{Q_{TBM} \left(\text{unfavourable joint orientation}\right)}{Q_{TBM} \left(\text{favourable joint orientation}\right)} = \frac{60}{10} \times \frac{1.4}{0.3} = 28$$

This will cause penetration rate PR to vary in accordance with Equation 11 (PR $\approx$ $5 \times Q_{TBM}^{-1/5}$) as the following examples demonstrate (see Table 13).

Sensitivity to orientation and to variable σ_c/I_{50} ratios is demonstrated by the above examples. If anything, greater sensitivity would be expected. However, the ultimate test is how the advance rate will vary in each case. From Equation 12, (AR $\approx 5 Q_{TBM}^{-1/5} \times T^{m}$) and use of Figure 45 (for estimation of $-m$) the following can be calculated (see Table 14).

Table 13. Examples of penetration rate estimates for anisotropic Q_{TBM} values.

QTBMU	=	0.28	2.8	28	280
PR (m/hr)	=	6.45	4.07	2.57	1.62
QTBMF	=	0.01	0.1	1.0	10
PR (m/hr)	=	(12.56)	(7.92)	5.00	3.15

Notes: 1) Unchanged cutter force F, and unchanged $SIGMA_{cm}/SIGMA_{tm}$ ratios are assumed in each of the above Q_{TBM} values. In practice, RQD_o and $SIGMA_{cm}$ and $SIGMA_{tm}$ values would tend to increase as Q_{TBM} increases. 2) Numbers in parenthesis imply operator intervention, and much reduced cutter loads, and consequently much more cautious PR values.

Table 14. Examples of advance rate estimates for anisotropic Q_{TBM} values.

Q_{TBM}	0.28	2.8	28	280	
m	−0.7	−0.5	−0.22	−0.17	
AR (m/hr)	(0.06)	0.15	0.60	0.53	(T = 720 hrs/1 month)
Q_{TBM}	0.01	0.1	1.0	10	
m	−0.7	−0.5	−0.22	−0.17	
AR (m/hr)	(0.13)	(0.30)	1.17	1.03	(T = 720 hrs/1 month)

Notes: 1) Values of $Q = Q_{TBMF}$ have been assumed for simplicity, in order to facilitate choice of gradient $(-)m$. In practice, $Q \neq Q_{TBM}$ in general. 2) As in Table 13, numbers in parentheses signify likely operator control of cutter loading due to 'too high' penetration rates.

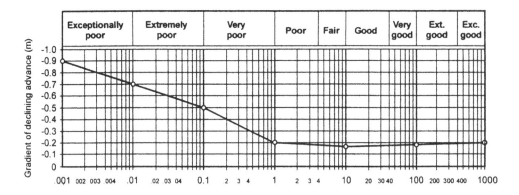

Figure 45. Preliminary estimate of declining advance rate gradient $(-)m$, as a function of Q-value.

The above examples demonstrate that the oriented Q_{TBM} method gives penetration rates and advance rates that are quite strongly orientation dependent. It is possible to imagine ratios of Q_{TBMU}/Q_{TBMF} as high as 100 in certain cases, which will increase the ratios of PR and AR for favourable and unfavourable directions to about 2.5, compared to about 2.0 in the above examples.

CHAPTER 23

Cutter wear and its effect on PR and AR

An additional fine tuning of Q_{TBM}, which also concerns machine-rock interaction (as does F/SIGMA), is the effect of limited cutter life in quartz-rich rock types. Two basic effects are involved. Firstly, there is an increase in the amount of ineffective rock crushing that occurs when a cutter gradually looses its original profile due to the abrasive nature of the rock. Secondly, more frequent stops are required for cutter-shift. An associated problem is caused in hard rock with marked individual joints, due to a reduced life for the cutter bearings. Increased numbers of joint sets increases the life of the cutter bearing (NTH 1994).

At the Svartisen hydroelectric project in Norway (World Tunnelling 1991), more than five times the volume of limestone was bored per cutter (215 m^3) compared to the quartzite (38 m^3). The mean result for the 9.3 km long tunnel was 87 m^3/cutter. Sixty percent of this tunnel was driven in mica schist.

The Siever's J-value (SJ) obtained from the miniature drill test shown in Figure 9, and the abrasion test (AVS = abrasion value steel) shown in the same figure have been used by NTH/NTNU engineers to define a cutter life index (CLI). Values of CLI for twelve rock types from Norwegian TBM tunnels are shown in Figure 11 (from Movinkel & Johannessen 1986). Cutter-life for different diameters of cutter and for the full range of CLI values (about 1 to 110) are also shown in Figure 11 (from NTH 1994).

The equation used by NTH to estimate the CLI can be rounded as follows: CLI ≈ 14(SJ/AVS)$^{0.4}$. This version includes an approximation of the power term ($^{0.4}$) which the original authors gave as 0.3847, which implies an accuracy that cannot exist in practice.

It is clear from the cutter life curves given in Figure 11 (bottom) that the lifetime (measured in the range of about 10 to 200 hrs) drops very rapidly for CLI values below about 20. This may therefore be a suitable value for normalising CLI, such that when the ratio 20/CLI is greater than 1, Q_{TBM} is increased (boring becomes difficult) and when 20/CLI is less than 1, Q_{TBM} is decreased, giving higher PR values. Trial and error shows that the power term $Q_{TBM}^{-1/5}$ (from Equations 11 and 12) gives an appropriate 'weighting' to CLI when PR is calculated from this 'fine-tuned' version of Q_{TBM}:

$$Q_{TBM} = \frac{RQD_o}{J_n} \times \frac{J_r}{J_a} \times \frac{J_w}{SRF} \times \frac{SIGMA}{F} \times \frac{20}{CLI} \tag{21}$$

Evaluating PR from Equation 11,

$$PR \approx 5Q_{TBM}^{-1/5} \tag{11}$$

we can readily see the effects of the CLI values. A typical 'mid-range' value of CLI = 20 is used in the first column of Table 15, while typical values for quartzite (CLI ≈ 4) and for shale (CLI = 80) are used in the second and third columns.

The need to shift cutters much more frequently in abrasive rocks and less frequently in porous, non-crystalline rocks means that the gradient (*m*) in Figures 34 and 35 will also be affected by CLI. The gradient will be steeper in the former case, and shallower in the latter, a similar effect to the delays cause by more support or no support respectively. It is therefore appropriate to also correct (*m*) values with CLI values, within the range experienced in practice. Once again trial and error shows that normalisation by 20/CLI is an appropriate starting point, and as in Q_{TBM} (Equation 11), the power term 0.2 is found to be useful. The following preliminary 'fine-tuned' gradient of declining advance rate is therefore as follows:

$$m = m_1 \left(\frac{20}{CLI} \right)^{0.2} \tag{22}$$

For the above examples of quartzite (CLI ≈ 4) and shale (CLI ≈ 80), Equation 22 gives correction factors of 1.38 and 0.76 respectively. These would have the effect of increasing and decreasing a typical gradient (*m* = –0.2) to values of (–)0.28 and (–)0.15 respectively. This lowest gradient corresponds to the best results plotted in Figure 34, when boring in predominantly stable limestone, while the higher gradient is similar to the 'extremely poor' trend line given in Figure 34, which has a gradient of (–)0.25, and would be typical for tunnels bored in predominantly abrasive, massive rocks.

Table 15. Effect of CLI on estimated PR values.

Q_{TBM} (before CLI term)	PR (m/hr) for following CLI values		
	CLI = 20	CLI = 4	CLI = 80
0.1	(7.9)	(5.7)	(10.5)
1	5.0	3.6	(6.6)
10	3.2	2.3	4.2
100	2.0	1.4	2.6
1000	1.3	0.9	1.7

Note: As in previous examples, PR values in parentheses signify likely operator adjustment.

An extreme, but real example can be given here to demonstrate the Q-Q_{TBM} method, and the influence of unfavourably *hard*, *massive* and *abrasive* rock. These three critical adjectives have a snowball effect on Q_{TBM} due to their implication of high σ_c, high Q and low CLI values. The low starting point for gradient (m) – there will be little need of support – is prejudiced by the low CLI value, as the gradient m is increased by all the cutter changes required.

Example 1
Massive, very hard, very abrasive quartzite, virtually unjointed over many tens of meters. Stresses are also high, but rock bursting or stress-slabbing is hardly experienced due to the extremely high uniaxial strength of 300 MPa.

$$Q_o \approx \frac{100}{1} \times \frac{4}{0.75} \times \frac{1}{0.5} \approx 1070$$

$$Q_c = Q \times \frac{\sigma_c}{100} \approx 3200$$

$$\text{SIGMA}_{cm} = 5\gamma Q_c^{1/3} = 5 \times 2.65 \times 3200^{1/3} \approx 195 \text{ MPa (from Equation 7)}$$

The TBM of 8 m diameter was installed 10 years ago by contractor No. 1 and is under-powered. Assume $F = 20$ tnf. Cutter life index ≈ 4, quartz content $\approx 85\%$

$$Q_{TBM} \approx \frac{100}{1} \times \frac{4}{0.75} \times \frac{1}{0.5} \times \frac{195}{20} \times \frac{20}{4} \text{ (from Equation 21)}$$

$$\approx 52,000$$

$$PR \approx 5 \times 52,000^{-0.2} \approx 0.57 \text{ m/hr (from Equation 11)}$$

From this unfavourable PR value, we can now estimate AR for 24, 168, 720 and 8760 hrs (nominal day, week, month, year).

$$AR \approx 5Q_{TBM}^{-1/5}T^m \ (= PR \times T^m) \text{ (from Equation 12)}$$

For

$$Q \approx Q_o \approx 1000, \ m \approx (-)0.21 \text{ (Fig. 45)}$$

$$m \approx (-)0.21 \left(\frac{20}{4}\right)^{0.2} \text{ (from Equation 22)}$$

$$\approx (-)0.29$$

$$\begin{aligned}
AR_{24} &\approx 0.57 \times 24^{-0.29} &\approx 0.23 \text{ m/hr (daily average)} \\
AR_{168} &\approx 0.57 \times 168^{-0.29} &\approx 0.13 \text{ m/hr (weekly average)} \\
AR_{720} &\approx 0.57 \times 720^{-0.29} &\approx 0.08 \text{ m/hr (monthly average)} \\
AR_{8760} &\approx 0.57 \times 8760^{-0.29} &\approx 0.04 \text{ m/hr (yearly average)}
\end{aligned}$$

In round figures, these 'extremely tough' boring conditions (refer to Fig. 44) suggest average advances of only 5.5 m/day, 22 m/week, 58 m/month and 350 m/year assuming three shifts, seven days per week and no vacation stops. Unfortunately such experiences are more or less a reflection of reality in this case. In fact even slower rates than this are experienced, and we will return to this example later when considering the adverse effects of high rock stresses.

Example 2

Sheared talcy phyllites with one, two and locally three joint sets, slickensided. Assume $\sigma_c \approx 10$ MPa, $\beta° = 30°$, $I_{50} \approx 0.01$ MPa (CLI = 50). In the worst case RQD = RQD_o = 10.

$$Q_o \approx \frac{10}{9} \times \frac{0.5}{4} \times \frac{1}{10} \approx 0.01$$

$$Q_t = 0.01 \times \frac{0.01}{4} \approx 0.000025$$

$$SIGMA_{tm} = 5\gamma Q_t^{1/3} \approx 5 \times 2.95 \times 0.000025^{1/3} \approx 0.4 \text{ MPa} \quad \text{(Equation 19)}$$

Insufficient stand-up time with the 5 m of no-support span (Table 10, Fig. 42) causes caving before support can be installed. (Conveyor produces more muck than the penetration of the TBM). We will first assume that very low cutter loads (i.e. $F \approx 1$ tnf) are needed to penetrate the phyllite, though torque is high. (In fact, as will be seen, this cutter force is unnecessarily high and the operator must intervene again.)

$$Q_{TBM} \approx \frac{10}{9} \times \frac{0.5}{4} \times \frac{1}{10} \times \frac{0.6}{1} \times \frac{20}{50} \approx 0.0024 \quad \text{(Equation 21)}$$

PR $\approx 5 \times 0.0024^{-0.2} \approx 16.7$ m/hr (theoretical, operator must intervene due to stability problems).

For $Q \approx Q_o \approx 0.01$, $m_1 \approx (-)0.70$ (Fig. 45). With CLI = 50 (favourable) the gradient $(-)m$ is reduced significantly:

$$m \approx m_1 \left(\frac{20}{CLI} \right)^{0.2} \approx (-)0.58$$

We will now estimate AR for 24, 168, 720, 8760 hrs (nominal day, week, month, year) using Equation 12:

$$AR \approx 5Q_{\mathrm{TBM}}^{-1/5}T^m \ (=\mathrm{PR}\times T^m)$$

We must assume that the operator intervenes, and reduces PR to well below the theoretical 16.7 m/hr estimated above. We can make the assumption that $\mathrm{PR}_{\mathrm{operator}} = 2$ m/hr, allowing time for support. In that case:

$$
\begin{aligned}
AR_{24} &\approx 2.0 \times 24^{-0.58} &\approx 0.32 \text{ m/hr (daily average)}\\
AR_{168} &\approx 2.0 \times 168^{-0.58} &\approx 0.10 \text{ m/hr (weekly average)}\\
AR_{720} &\approx 2.0 \times 720^{-0.58} &\approx 0.04 \text{ m/hr (monthly average)}\\
AR_{8760} &\approx 2.0 \times 8760^{-0.58} &\approx 0.005 \text{ m/hr (yearly average)}
\end{aligned}
$$

In round figures, these 'extremely/very poor' stability conditions (refer Fig. 44) suggest average advances of only 8 m/day, 17 m/week, 29 m/month and 44 m/year. In fact these extremes have been more or less experienced in practice (in the same tunnel as Example 1).

CHAPTER 24

Effect of porosity and quartz content on gradient m and PR

Two further aspects have to be considered when cutter life is to be assessed. Johannessen et al. (NTH 1994) have given a graphic description of the types of wear experienced by cutters. Hard abrasive rock has a general tendency to blunt the cutters, with more side abrasion in the less hard rock types. When low resistance to penetration and low abrasion are combined, the cutter slowly loses material from all surfaces. If penetration is particularly easy, more material flows past the sides of the cutter, more heat is generated and self-sharpening occurs. Quartz content in addition to CLI values, therefore has influence, together with porosity.

Porosity increase, which is usually linked to lower uniaxial strength and lower density is partly taken account of already during the estimation of $SIGMA_{cm}$ and $SIGMA_{tm}$ (Equations 7 and 19). This will cause a reduced value of Q_{TBM} and easier penetration. At the same time, cutter wear may be enhanced by the high rate of advance and there will be a certain negative influence on gradient $(-)m$.

It is not known at present exactly how much reduction this will cause, but the following simple modifications to Equation 22 are suggested at this stage. They appear to give realistic gradients over a wide range of conditions that agree in general terms with the case records from which Figure 34 trend lines were derived.

$$m \approx m_1 \left(\frac{20}{CLI} \right)^{0.15} \left(\frac{q}{20} \right)^{0.10} \left(\frac{n}{2} \right)^{0.05} \tag{23}$$

where n = porosity in %, and q = quartz content in %.

To avoid problems with zero and to retain simplicity, Equation 23 should be used with $q \geq 0.5\%$ and $n \geq 0.5\%$. When values of q and n are lower than these values, simply use 0.5%.

It may be observed from Equation 23 that a soft, porous, but abrasive quartz rich rock will have conflicting effects on penetration rate. Firstly, it will cause lower values of $SIGMA_{cm}$ and $SIGMA_{tm}$ to be estimated, since σ_c and I_{50} and γ will each be lower. However, the ratio 20/CLI in Equation 21 will be greater than unity, and Q_{TBM} will increase somewhat. Nevertheless PR will be quite fast.

However, when advance rate is considered, which requires gradient $(-)m$ from Equation 23 (and Table 11), the gradient will be found to increase due to the higher ratio of 20/CLI, due to (q/20) probably >1, and due to (n/2) far greater than unity.

Example

With CLI = 8, q = 40%, n = 20%, a typical gradient m_1 = (–)0.2 would be changed as follows: $m \approx$ (–)0.2 × 1.15 × 1.07 × 1.12 $\approx$ (–)0.28. This will mean that despite (and partly because of) high PR, lower AR-values than expected will probably be experienced, due to extra cutter wear problems of the 'self-sharpening' variety. 'Too much' abrasive material will be flowing past the sides of the cutter, due to the high penetration rates that can be achieved with ease, but which are paid for by reduced cutter life.

CHAPTER 25

Tunnel size effects

Unfortunately, it is not so frequently that a large diameter TBM follows a small diameter TBM through the same rock, without the 'benefit' of the reduced volume to excavate, caused by the pilot tunnel preceding it. If there is a 'normal' distribution of rock quality, the large TBM will encounter a longer total length of rock requiring temporary (and permanent) support than the smaller TBM. This will tend to increase gradient $(-)m$ in these zones, and tend to give a slightly higher overall average gradient as a result.

The 'no-support boundary' in the Q-system rock support diagram and its influence on the 'central threshold' Q-values (where TBM Q-values are higher than drill-and-blast Q-values) will be discussed later in this book. The need for more support in the large TBM tunnel can be related to the Q-value. In very good quality rock, no support will be required in either the large or small TBM tunnels.

There are however, complicating factors in the above intuitive treatment of TBM size and support needs. Firstly, it will usually be much more efficient to provide support in the large diameter tunnel, despite the larger quantities required (i.e. more bolts, heavier steel arches, more volume of shotcrete, larger and heavier PC element liner units).

So although more overbreak occurs in the larger tunnel (e.g. Scesi & Papini 1997), the necessary support can often be placed more efficiently using more drills and more working platforms behind the cutter-head. Indirect evidence for this is given by the TARP project in the Chicago area. Here a unique set of results from 27 TBM ranging from 2 m to 10.8 m in diameter are available, giving experience from a total of some 150 km of tunnels. In general, these TBM were each boring in a rather consistent bedrock consisting of Silurian dolomite ($\sigma_c \approx 120\text{-}230$ MPa) with widely spaced jointing.

Dalton et al. (1993), showed that under these uniform and favourable conditions, there was a linear *increase* in advance rate with size (see Fig. 46). The five-fold increase in diameter gave a 20% increase in monthly rate of advance (AR $\approx$ 0.8 m/hr with 2 m TBM and about 0.9 m/hr with 10.8 m TBM for best month, and AR $\approx$ 0.4 m/hr to 0.5 m/hr for average month). Naturally, the overall project design philosophy and machine characteristics, including average rated cutter loads could influence such results in either direction.

An extensive review of penetration rates achieved in hard rocks ($\sigma_c \geq 70$ MPa)

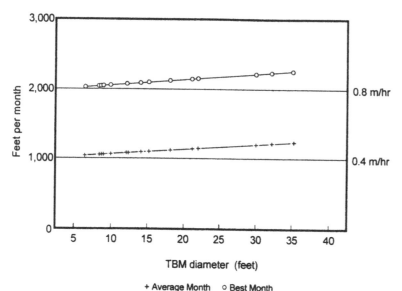

Figure 46. Linear effect of TBM diameter on advance rate for numerous TARP tunnels in the Chicago area (Dalton et al. 1993).

in 83 TBM tunnels with diameters ranging from about 2.2 to 12 m, suggests in general no significant scale effect on PR (Stevensen 1999). There is wide scatter for all diameters between 2 and 8 m. In the last 20 years (since 1980) cutter thrusts are seen to have increased from about 10 to 20 tnf (lower bound trend line) or from about 25 to 35 tnf (upper bound trend line). These increases accompanying the manufacture and use of larger machines, may be partly responsible for the apparent lack of scale effect on PR.

Although large tunnels may be driven faster than small tunnels in similar good rock conditions, more delays occur in the larger tunnel if the rock is consistently poor. Therefore a normalised tunnel diameter (D) of 5 m has been selected to slightly modify the gradient $(-)m$. (Q_{TBM} is already partly adjusted for tunnel size by the use of average rated cutter force.) The resulting 'fine-tuned' gradient $(-)m$ is estimated as follows:

$$m \approx m_1 \left(\frac{D}{5}\right)^{0.20} \left(\frac{20}{CLI}\right)^{0.15} \left(\frac{q}{20}\right)^{0.10} \left(\frac{n}{2}\right)^{0.05} \tag{24}$$

To avoid zero problems, yet retain simplicity, both (q) and (n) should again be set to $\geq 0.5\%$.

The TARP results reviewed here (AR $\approx$ 0.4 m/hr with D = 2 m, AR $\approx$ 0.5 m with D = 10.8 m) can be used to demonstrate that the gradient $(-)m$ must have reduced with the larger diameters, if we assume that Q_{TBM} is constant (same cutter forces, same rock masses). Equation 12, evaluated with T = 500 hrs (assumed, per month) and a reference $m = -0.20$, gives $Q_{TBM} = 610$ when AR = 0.4 m/hr, and $Q_{TBM} = 200$ when AR = 0.5 m/hr (see also Fig. 44). If however we assume $Q_{TBM} = 610$ is 'correct', then the gradient $(-)m$ must have changed from $(-)0.20$ to $(-)0.16$ as the larger machines were used. A low gradient (good efficiency) but relatively poor PR value in fact was noted when analysing TARP results in the format shown in Figure 35.

For a more typical range of rock qualities, the assumption has been made in Equation 24 that the efficiencies of the larger machines will be outweighed by more delays due to support requirements increasing as diameter increases. In Equation 24, if we set $m_1 = (-)0.2$, CLI = 20, $q = 20\%$, and $n = 2\%$, the $D = 2$ m to $D = 10.8$ m, TARP range of diameters would 'normally' cause an increase in gradient from $(-)0.17$ to $(-)0.2$, to $(-)0.23$ as one progresses from 2 m to 5 m to 10.8 m diameter. This designed correction for diameter in equation 24 will give slower advance rates as diameter increases, but not a dramatic increase. This seems to be in line with expectation. When rock quality is consistently poor (larger $(-)m_1$ from Table 11), the correction for diameter will be more marked, as desired.

CHAPTER 26

Boring in exceptionally tough, high-stress conditions

In Figure 44, the adjectives tough, very tough, exceptionally tough and so on are implied when the Q_{TBM} value exceeds 100, 1000 and 10,000. The Q_{TBM} range does not stop at 10,000 however, a fact that is suitably demonstrated by Table 16.

A useful case record for extending the Q_{TBM} model to its limits, and from which the obvious importance of high stress can be realised, is the aborted Star Mine, Idaho case described by Hendricks (1969), and more recently reviewed by Tarkoy & Marconi (1991). The exceptionally poor PR of 'less than' 0.12 m/hr is shown in Figure 35 marked 'high stress' (lower left-hand side of figure).

The Revette quartzite at Star Mine was described as hard, brittle, intensely fractured and highly stressed. According to Tarkoy & Marconi (1991), the underpowered machine could achieve only about 1 tnf/cutter, and the attempts to drive mining drifts by TBM were abandoned after 125 m. The depth was an exceptional 2250 m. Other cases of exceptionally high stresses are described where TBM have ground to a halt when the rock was 'as hard as steel'.

At Mont Vernis in the French Alps a short section of the tunnel gave PR as low as 0.18 m, and in fact it was the delay of re-gripping that gave enough stress-induced fracturing of the face to allow initial boring progress in each cycle. Behind the machine the rock popped and burst, giving 8 to 10 cm of overbreak. At a hydroelectric project in Columbia, using 20 tnf/cutter, the very hard quartzite (99% quartz content) allowed only 10 m^3 per cutter to be excavated.

There are records of lavas with uniaxial compressive strengths in excess of 600 MPa in South African gold mines. Although the slow penetration rates may be predictable, deep mining experiences in South Africa suggest that cutter life pre-

Table 16. Estimates of Q_{TBM} for lines WR, 1, 2, 3 and 4 in Figure 34.

From Figure 34	PR m/hr	Q_{TBM}
WR ('world record')	10	≈ 0.03
Line 1 ('good')	5	≈ 1
Line 2 ('fair')	3	≈ 13
Line 3 ('poor')	2	≈ 98
Line 4 ('extr. poor')	1	≈ 3125
Exceptional case	0.1	$\approx 3.1 \times 10^8$

diction and overall machine utilisation are hard to predict due to the unfavourable interaction of high stresses and face and wall stability. Large slabs of loosened rock create significant problems both for the cutters, cutter head and shield, and subsequently also for the grippers, which may not reach the sidewalls even at full stroke, if 'dog-earing' is of major proportions (Pickering et al. 1999).

Returning to the Star Mine case record, one can make the preliminary assumption that the rock mass was indeed 'highly fractured' (rather than stress-induced fractured) and see how far one progresses with the Q, Q_{TBM} approach. The following approximate Q-parameters will be assessed:

$$Q_o \approx \frac{30}{12} \times \frac{2}{1} \times \frac{0.66}{0.5} \approx 6.6$$

Assume $\sigma_c \approx 350$ MPa, $\gamma = 2.7$ gm/cm^3, CLI = 3 (lower 25% of NTH 1994 results shown in Fig. 11), RQD$_o$ = RQD. $Q_c \approx 6.6 \times 350/100 \approx 23$, and from Equation 7, SIGMA$_{cm} \approx 5 \times 2.7 \times 23^{1/3} \approx 38$ MPa.

From Equation 21, one can therefore write:

$$Q_{TBM} \approx \frac{30}{12} \times \frac{2}{1} \times \frac{0.6}{0.5} \times \frac{38}{1.0} \times \frac{20}{3} \approx 1680$$

This value of Q_{TBM}, which lies only in the 'very tough' area in Figure 44, is clearly inadequate to explain such a low PR as 0.12 m/hr. In fact, Equation 11 would suggest PR $\approx$ 1.1 m/hr, based on Q_{TBM} = 1680.

There are several possibilities for improving the estimates of very low PR values. One is that the exceptional depth (2.25 km) imparts such a high shear strength to the joints and fractures that they are effectively 'locked' in relation to the inadequate cutter forces of reportedly, only $\approx$ 1 tnf. Alternatively, one could make the assumption that the rock is actually rather massive until stress-fractured at the periphery of each mining excavation.

The 'massive rock' assumption can be tested by mobilising all the appropriate Q-parameters that describe unjointed, healed conditions (see Q-ratings in the Appendix).

$$Q\,(\text{intact}) \approx \frac{100}{0.5} \times \frac{5}{0.75} \times \frac{1.0}{0.5} \approx 2667$$

$$Q_c \approx 2670 \times \frac{350}{100} \approx 9333$$

$$\text{SIGMA}_{cm} \approx 5 \times 2.7 \times 9333^{1/3} = 284 \text{ MPa}$$

From Equation 21 one can therefore write:

$$Q_{TBM} \approx \frac{100}{0.5} \times \frac{5}{0.75} \times \frac{1}{0.5} \times \frac{284}{1.0} \times \frac{20}{3} \approx 5,048,000$$

With this exceptionally high Q_{TBM} value, equation 11 now predicts a significantly lower and more realistic PR $\approx$ 0.22 m/hr. However, stress as a factor should probably also be considered. In order to discover what 'stress correction' one perhaps requires, the exceptional PR = 0.12 m/hr from the Star Mine can be used to back-calculate Q_{TBM} using Equation 25 (this is a simple rearrangement of Equation 11).

$$Q_{TBM} \approx \left(\frac{5}{PR} \right)^5 \tag{25}$$

In the same way Equation 12 can be rearranged for back-analysis of performance:

$$Q_{TBM} \approx \left(\frac{5T^m}{AR} \right)^5 \tag{26}$$

For Star Mine, with PR = 0.12 m/hr, Equation 25 suggests:

$$Q_{TBM} \approx \left(\frac{5}{0.12} \right)^5 \approx 125,587,000$$

The ratio of these two large Q_{TBM} values (125,587,000/5,048,000) is about 25. An overburden of 2.25 km as at Star Mine, will theoretically generate a vertical stress in the region of 60 MPa if $\gamma \approx 2.7$ gm/cm^3. An obvious, but perhaps too simple measure would be to add σ_v (vertical stress) to Equation 21. In fact, the biaxial stress state on the face of the tunnel may be the chief source of penetration resistance caused by stress, so a formulation (normalisation) using the average induced biaxial stress σ_θ would be more valid.

The magnitude of σ_θ (average biaxial stress on tunnel face) will probably be of the same order of magnitude as the tangential stresses around the cylindrical walls of the tunnel. If we assume no penetration difficulties before a normal depth of about 100 m is reached, then a normalised value $\sigma_\theta/5$ could be tested, which in the case of the Star Mine (if stresses were approximately isotropic) could be expected to give a value $\sigma_\theta/5 \approx 120/5 \approx 24$, based on $\sigma_v \approx 60$ MPa.

Such speculation allows one to close in on the Q_{TBM} 'goal' of 1.25×10^8 given above. However, it may be more correct to assume that the quartzite will have had a bedded structure (possibly even a tectonically induced set of joints if horizontal stresses are anisotropic, e.g. the Lucky Friday mine in Idaho, where steeply dipping bedding and steeply dipping joints probably sub-parallel to σ_H were logged in a 1.6 km deep shaft in quartzite (Barton & Bakhtar 1983).

It may therefore be wise to re-evaluate our Q(intact) value, adding the possible reality of bedding joints and a set of tectonic joints i.e. J_n increases from 0.5 to 2 or 4, and J_r/J_a possibly changes from 5/0.75 to (2 or 3)/1, assuming that the joints are not perceived as completely 'locked' by the high stresses. This will be an area for engineering judgement.

When establishing as realistic a gradient $(-)m$ of declining advance rate as possible in Chapter 24 (Equation 23) and in Chapter 25 (Equation 24), the quartz content and porosity were included with cutter life index CLI (NTH 1994) to 'fine-tune' the value of $(-)m$. Improved realism for Q_{TBM} estimation may be achieved if quartz content is also included in a normalised manner. A 'final' Q_{TBM} formulation may therefore be as follows:

$$Q_{TBM} \approx \frac{RQD_o}{J_n} \times \frac{J_r}{J_a} \times \frac{J_w}{SRF} \times \frac{SIGMA}{F} \times \frac{20}{CLI} \times \frac{q}{20} \times \frac{\sigma_\theta}{5} \qquad (27)$$

Returning to the Star Mine, and allowing for some jointing, the following can perhaps be assumed:

$$Q_o \approx \frac{100}{2 \text{ to } 4} \times \frac{2 \text{ to } 3}{1} \times \frac{1}{0.5} \approx 200 \quad (\text{range} \approx 100 \text{ to } 300)$$

$$Q_c \approx 700 \quad (\text{range } 350 \text{ to } 1050), \quad SIGMA_{cm} \approx 120 \text{ MPa}$$
$$(\text{range } 95 \text{ to } 140 \text{ MPa})$$

$$Q_{TBM} \approx \frac{100}{2 \text{ to } 4} \times \frac{2 \text{ to } 3}{1} \times \frac{1}{0.5} \times \frac{120}{1} \times \frac{20}{3} \times \frac{100}{20} \times \frac{120}{5} \approx 16,000,000$$
$$(\text{range } 10,000,000 \text{ to } 29,000,000)$$

From Equation 11, PR ≈ 0.18 m/hr (range 0.20 to 0.16 m/hr).

If there is still a desire to be even closer to the recorded PR $= 0.12$ m/hr, then an even less favourable value of CLI $= 2$ could be used, and a tectonic stress giving a higher value of σ_θ could be assumed. This would readily increase Q_{TBM} to about 10^8, giving the desired result.

CHAPTER 27

Revisiting cutter force effects

Before leaving Equation 27 as the 'final version' for estimating Q_{TBM}, the possible limits of Q_{TBM} can again be explored, this time considering variations in cutter load. We must make a reasonable assumption that $F = 1$ tnf will be the limiting case (e.g. Star Mine). However, through faulted rock even lower values than 1 tnf could be used. Obviously today a minimum rated F value probably in excess of 10 tnf would be realistic.

For this example, we will assume the toughest likely quartzite in a completely intact, massive rock mass in a deep level gold mine (say 3500 m deep) outside the mining-affected fractured rock. For such conditions we may estimate:

$$Q_{TBM}(max) \approx \frac{100}{0.5} \times \frac{5}{0.75} \times \frac{1}{0.5} \times \frac{300}{1} \times \frac{20}{2} \times \frac{100}{20} \times \frac{200}{5} \approx 1.6 \times 10^9$$

From equation 11, the predicted minimum PR value would be 0.07 m/hr. If the cutter force was raised to 15 tnf and then to 30 tnf, with all other factors unchanged, Q_{TBM} would be 'reduced' to 1.1×10^8 and 5.3×10^7 respectively, giving predicted PR values of 0.12 m/hr and 0.14 m/hr respectively. This prediction (which needs improvement) would not auger well for using TBM in deep mines unless the stress-induced fracturing was contributing, as at Mont Cernis, between cutter regrip (Tarkoy & Marconi 1991).

Between the approximate limits of Q_{TBM} of 10^{-4} to 10^9, Equation 11 suggests a theoretical 500:1 range of PR. However, this is entirely hypothetical because a PR value of 30 m/hr (a so-called 'geotechnically correct' value) is impossible unless the exceptionally poor conditions were stabilised, in which case Q_{TBM} would be much larger than 10^{-4} anyway. In practice, a 100:1 range of PR could be expected today, i.e. 10 m/hr down to 0.1 m/hr. TBM design changes in the future will tend to move both limits upwards. It will however be more important to narrow today's AR range of about 7 m/hr (the world record for 1 day at the Australian Blue Mountains project) to the minimum of about 0.003 m/hr (e.g. at Pinglin) by suitable probing, sonic logging and pre-treatment methods.

The preceding attempts to 'explain' the extremely low penetration rates in the Star Mine (0.12 m/hr), by incorporating logical corrections to Q_{TBM} for quartz content [normalised ratio ($q/20$) and biaxial stress ($\sigma_\theta/5$)], as shown in Equation 27 have probably been useful in the long term. However, they have masked an important omission (actually an error) in Q_{TBM}, which will now be addressed.

The ratio *F*/SIGMA which was originally located in place of the *Q*-value SRF term in Equation 6, has been inverted (i.e. SIGMA/*F*) in the improved format shown in Equation 27. This ratio expresses the possibility of reduced penetration rate with increased cutter force if the rock is too hard (by reducing the magnitude of Q_{TBM}). It also allows increased penetration rate if *F* is increased and SIGMA remains constant. Because our attention has so far been focussed on geotechnical processes that slow or speed up penetration, insufficient attention has been paid to the effect of the negative quintuple power term $Q_{TBM}^{-0.2}$ in Equation 11. In fact, the present proportionality between Q_{TBM} and the cutter force *F* is already too weak a coupling, and the importance of cutter force is drastically reduced by this negative power term. This limitation was recognised when exploring the limits of Q_{TBM} using widely varying cutter forces, in place of typical values such as 15 or 20 or 25 tnf, as earlier.

Reference to Figure 15 confirms the very powerful potential effect of cutter force if the strength of the rock is held constant. Something resembling a quadratic increase in penetration rate with increased cutter force (all other factors being held constant) is indicated. The good quality, hard rock IMF class 1 rock mass of Grandori et al. (1995a) (maybe $Q \approx 50$) is least affected by cutter force, while the slightly weathered IMF class 2 is more strongly affected (see Fig. 15). A rather marked change of gradient is seen when *F* is increased above 20 tnf. Similar sharply curved PR-*F* curves have been shown by Nelson (1993), for high strength rock and low strength rock respectively.

Trial and error, and simple mathematics, show that a quadratic law such as PR $\propto F^2$ (nine times the rate of penetration for a three-fold increase in *F*) will need tenth power formulation in Q_{TBM}, if this parameter is subsequently factored as in Equation 11 ($Q_{TBM}^{-0.2}$). However, wishing to maintain the simplicity of Equation 11, and a middle value of $F = 20$ tnf, means that the normalised ratio $F^{10}/20^9$ needs to be used in a modified version of Equation 27. Thus, to obtain a real quadratic PR $\propto F^2$ relation in Equation 11, we need to write the following:

$$Q_{TBM} \approx \frac{RQD_o}{J_n} \times \frac{J_r}{J_a} \times \frac{J_w}{SRF} \times \frac{SIGMA}{F^{10}/20^9} \times \frac{20}{CLI} \times \frac{q}{20} \times \frac{\sigma_\theta}{5} \qquad (28)$$

This format means that when $F = 20$ tnf, SIGMA will be divided by 20 tnf. Lower cutter forces will have strongly diminished effect (Q_{TBM} increases, PR reduces) and the opposite will occur when $F > 20$ tnf. In each case the result of $F^{10}/20^9$ is compared to SIGMA, the relevant strength of the rock mass. The ratio $(20^9 \times SIGMA)/F^{10}$ has an 'exaggerated' effect on Q_{TBM}, but gives the 'correct' quadratic level of influence on PR, when using Equation 11:

$$PR \approx 5Q_{TBM}^{-1/5}$$

The five curves of PR versus *F* given in Figure 47 have been evaluated for five

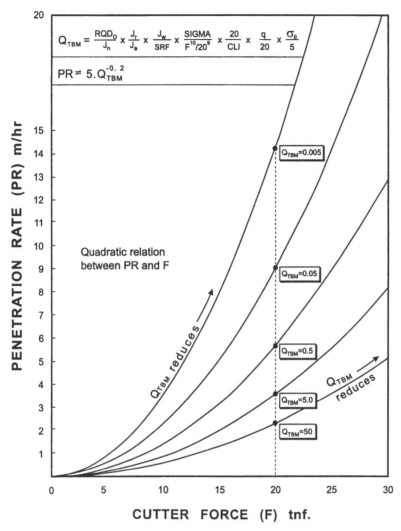

Figure 47. Penetration rate versus cutter force illustrating the quadratic relationship assumed. Five example values of Q_{TBM} (five curves) and their variation with F, following Equation 28.

sets of Q_{TBM} values. Since Q_{TBM} varies (strongly) with cutter force F, only the reference values of Q_{TBM} at $F = 20$ tnf (0.005, 0.05, 0.5, 5 and 50) are labelled. These represent extreme faulted conditions, up through ideal Q_{TBM} values (see Fig. 44) and continue into 'fair' conditions when $Q_{TBM} = 50$. Table 17 is the source of the data in Figure 47.

The final version of Figure 44, with Q_{TBM} given by the complete set of eleven parameters (Equation 28) can be found in the Appendix – Section A2. It is located together with the Q-system parameter ratings given in the Appendix – Section A1 (i.e. ratings for the first six parameters constituting both Q_o and Q_{TBM}). The location of this 'Q_{TBM}' figure in the appendix is deliberate. It will make any improvements (i.e. further fine tuning of Q_{TBM}) easier in any subsequent edition.

Table 17. Example of PR versus F for five orders of magnitude of Q_{TBM}.

F (tnf)	$F^{10}/20^9$	Q_{TBM} values				
5	0.000019	5263	52,632	5.26×10^5	5.26×10^6	5.26×10^7
10	0.0195	5.1	51.2	512	5120	51,203
15	1.126	0.089	0.89	8.89	88.9	889
20	20	0.005	0.05	0.5	5	50
25	186	0.00054	0.0054	0.054	0.54	5.37
30	1153	8.7×10^{-5}	8.7×10^{-4}	0.0087	0.087	0.87
F tnf		PR $= 5 \times Q_{TBM}^{-1/5}$ (m/hr)				
5	–	0.90	0.57	0.36	0.23	0.14
10	–	3.6	2.3	1.4	0.9	0.6
15	–	8.1	5.1	3.2	2.0	1.3
20	–	(14.4)	9.1	5.7	3.6	2.3
25	–	(22.5)	(14.2)	9.0	5.7	3.6
30	–	(32.4)	(20.7)	(12.9)	8.1	5.1

Note: 'geotechnically correct' but extreme PR values given in parentheses will of course be reduced by the operator. If the faulted rock is pre-treated (as it should be), the Q-value will increase and Q_{TBM} will also increase thereby reducing PR automatically.

CHAPTER 28

Predicting advance rates in faulted rock

The difficulties caused by faulted rock were discussed in earlier chapters, and are well illustrated by the examples given by Robbins (1982) and Grandori et al. (1995b) (see Figs 5 and 37). Faulted rock presents a difficulty not only to the TBM but also to the engineering geologist, because the operator has to 'over-rule' the natural tendency of the machine to penetrate faulted rock faster (at first), until either the cutterhead gets blocked, or the buckets or conveyor get overloaded, or a new set of problems arise when the grippers enter the same zone. This makes prediction of delays more difficult, because an unstable situation can develop in many different negative directions, and is decision-dependent or operator-dependent.

Figure 44 showed deliberately stippled lines for $Q_{TBM} < 1$, because TBM operator intervention is assumed when PR becomes too high. In other words, the average rated cutter loads of say 30 tnf for a high powered HP-TBM will be greatly reduced by the operator as the cutterhead registers less resistance to penetration. There will however usually be more torque demand due to greater cohesional and frictional resistance around the cutterhead which is no longer 'at-arm's-length' (cutter rim width) from the previously hard-rock tunnel face.

Equation 11 can be regarded as 'geotechnically correct' in the sense that it blindly assumes increased PR as Q_{TBM} reduces. Already at $Q_{TBM} = 0.001$, it predicts an 'impossible' 19.9 m/hr, and at $Q_{TBM} = 10^{-4}$ (extremely to exceptionally problematic ground) it predicts a hypothetical 31.5 m/hr. If stability could be assured (which it cannot at present), such rates could perhaps be achieved in theory – with four times the present mucking capacity!

Choice of relevant declining advance gradients for faulted rock was attempted in Table 11, on the basis of the 'blue' (bad ground) case records from which Figure 34 data trends were derived. In Figure 44, the gradients $m = -0.5$, -0.7 and -0.9 have been assumed at the lower end of the Q-value scale, and they have, thus far, been combined with the 'geotechnically correct' penetration rate curve (Equation 11) which ascends to impossible heights to the left hand side of the figure. We therefore have three theoretical (and over-optimistic) AR curves in this region.

In practice, when conditions are very bad, or when a fault has been partly pretreated, the operator might advance at a PR of 1 or 2 m/hr (when not actually stopped) so that steel rib erection or eventual bolting or shotcreting can be performed without getting unsupported too far behind the finger shield area. This means that average advance rates through a fault zone will sink, even in relation to the three stippled AR curves in Figure 44, which already show low rates of advance. According to our empirical model, this is because the cutter force (F) given in Equation 28 will be reduced to some low value by the operator, and Q_{TBM} will increase (see Fig. 47).

In order to give a feel for Q_{TBM} values at the faulted end of the twelve orders of magnitude Q_{TBM} scale, the following extreme, faulted conditions will be assumed. Q parameter ratings are estimated from Barton & Grimstad (1994) (see Appendix).

Example
TBM diameter = 3.5 m, faulted crushed sandstone, multiple clay gouges, depth = 600 m, σ_θ(est.) $\approx$ 15 MPa, (due to 'plastic' state of stress) high water pressure on one side of the fault. (σ_c sst. = 25 MPa, γ = 2.2 gm/cm^3, CLI (of sandstone blocks) $\approx$15, q = 25%, $n \approx$ 10%.

$$Q_o \approx \frac{10}{20} \times \frac{1}{8} \times \frac{0.33}{10} \approx 0.002 \quad \text{(exceptionally poor)}$$

Gradient $(-)m_1$ from Table 11, Figure 45 $\approx$ –0.83.
From Equation 24,

$$m \approx -0.83 \left(\frac{3.5}{5}\right)^{0.20} \left(\frac{20}{15}\right)^{0.15} \left(\frac{25}{20}\right)^{0.10} \left(\frac{10}{2}\right)^{0.05}$$

$$\approx -0.83 \times 0.93 \times 1.04 \times 1.02 \times 1.08$$

$$\approx -0.87$$

$$Q_c \approx 0.002 \times \frac{25}{100} = 0.0005$$

$$\text{SIGMA}_{cm} \approx 5 \times 2.2 \times 0.0005^{1/3} \approx 0.9 \text{ MPa}$$

If we assume that the cutter load is reduced to a nominal 1.5 tnf, then the value of Q_{TBM} can be estimated as follows (from Equation 28):

$$Q_{TBM} \approx \frac{10}{20} \times \frac{1}{8} \times \frac{0.33}{10} \times \frac{0.9}{1.5^{10}/20^9} \times \frac{20}{15} \times \frac{25}{20} \times \frac{15}{5} \approx 8.24 \times 10^7$$

The theoretical 'geotechnically correct' PR value from Equation 11 is in this case a much lower value, due to the low cutter force assumed:

PR $\approx$ 0.13 m/hr

and such a low rate would in principle be compatible with the heavy support requirements. However, experience and our empirical model, suggest almost infinite delays in such a fault zone unless the Q-value and gradient $(-)m$ can be improved by pre-treatment.

The time (T) required to drive the tunnel a length L (i.e. through a given rock class or through this fault zone) is given by the advance rate, i.e. $T = L/\mathrm{AR}$. As shown earlier, the time taken can therefore be quantified as follows (Equation 16):

$$T = \left(\frac{L}{\mathrm{PR}}\right)^{\frac{1}{1+m}}$$

If the fault zone in the above example is encountered for 5 m in the tunnelling direction, then the following, preliminary empirical estimate of time required to come through the fault will be as follows, based on the 'geotechnically correct' PR value derived above from an assumed reduced cutter load.

$$T \approx \left(\frac{5}{0.13}\right)^{\frac{1}{1-0.87}} \approx 38.46^{7.7} \approx 1.6 \times 10^{12} \text{ hrs}$$

This implies effective 'standstill' and emphasises the acute nature of fault zones and of the decisions made. The Pinglin tunnel in Taiwan has faulted rock with probably this level of extreme Q-value magnitude, and all three TBM's have been stopped as a result. If we put our trust in the fundamental nature of Equation 16, then there appear to be two possibilities for improving progress in an extreme fault zone. One practical approach will be to reduce the gradient $(-)m$ so that the component $1/(1+m)$ is reduced. This implies improving the quality of the fault zone (feasible) or removing the fault zone (obviously impossible). Increasing the cutter forces to make the fault width L (m) and PR (m/hr) numerically closer (thereby 'avoiding' problems with a big gradient $(-m)$ in Equation 16) is a very big risk because the fault zone has virtually no stand-up time (when $Q \approx 0.01$ to 0.001), so the machine is extremely likely to get stuck, or the tunnel will collapse if the machine does not get stuck. Going moderately slowly in an improved fault quality is the safest and perhaps only reliable approach.

Both in the case of drill-and-blast and TBM tunnels, it is imperative to improve the quality of the ground by drainage (increasing J_w) and by pre-injection (potentially improving many of the six Q-parameters). Otherwise the 'zero' stand-up time will cause collapse or a stuck machine. One then partly avoids the situation described by Equation 16, in which the component $[1/(1+m)]$ is far too large. Unless the Q-value is improved by pre-treatment, the situation becomes too sensitive to the ratio of fault length and penetration rate, in which operator decisions are far too important for comfort.

Most civil and mining engineers know from experience that correctly carried out pre-grouting reduces leakage, and that it seems to increase deformation modulus and probably shear strength, since tunnels that are pre-injected often show each of these physical characteristics. The need for support is obviously reduced, and tunnel deformation is reduced. Since tunnel deformation is closely linked to SPAN/Q (Barton et al. 1994) and support needs are linked directly to $Q^{-1/3}$, the inescapable conclusion (which would also be arrived at by seismic velocity monitoring and deformability testing) is that the effective Q-value has been increased by the pre-injection.

The Q-value is determined from RQD, the number of joint sets (J_n), the roughness (J_r) and degree of alteration (J_a) of the least favourable set, and from the water inflow (J_w) and stress/strength condition (SRF). A seismic velocity increase of 1.5 km/s from say 2.5 to 4.0 km/s in a wet, faulted zone ahead of a large tunnel, will imply that the Q-value has increased from about 0.1 to 3 as a result of the grouting. A reduction in Lugeon value is also implied, and the modulus of deformation may also be predicted to have increased (Barton 1999, 2000).

Are these changes possible to explain via changes in the six component Q-parameters? The answer is definitely yes, but the exact answer will always be unknown. We could speculate that approximately the following may occur in principle when a typical faulted zone is pre-injected ahead of the tunnel face.

1. RQD of 10% increases to about 30% due to grouting of the most prominent discontinuities that were most permeable.

2. J_n of 15 (four sets) is effectively reduced to about 9 (three sets) for the same reason as above.

3. J_r of 1.5 (rough, planar) changes to 2 (another set).

4. J_a of 6 (sand filled) changes to 1 (another set).

5. J_w of 0.5 (high pressure inflow) changes to 0.66 (small inflow). (Drainage time will also help here.)

6. SRF of 1 (unchanged). (In the case of a minor fault even SRF might be changed by grouting.)

We therefore have the following potential 'before' and 'after' scenarios:

$$Q_1 = \frac{10}{15} \times \frac{1.5}{6} \times \frac{0.5}{1} = 0.08 \quad \text{(extremely poor)}$$

$$Q_2 = \frac{30}{9} \times \frac{2}{1} \times \frac{0.66}{1} = 4.4 \quad \text{(fair)}$$

The effective Q-value has increased, which is broadly consistent with the expected increases in V_p and deformation modulus values, and with reduced Lugeon value and rock support needs. These coupled effects are discussed by Barton (1999, 2000). They are very important physical results for TBM tunnelling.

Advance rates through fault zones will be more predictable when the fault has been probed and pre-treated. When the Q-value is increased, Q_{TBM} increases and the gradient (–)*m* reduces (Table 11). The days spent on pre-treatment may pay off handsomely in more 'constant' advance rates. Anything to reduce (–)*m* – the gradient of deceleration (units of LT^{-2}) will be positive for TBM performance. In other words, anything to increase utilisation – especially the average utilisation – must be striven for, even if this means an apparent short-term reduction in utilisation due to grouting or pre-support.

The improvement in the *Q*-value achieved in the above example is not unrealistic, and in practice might range from an improvement of 10 to 100 times the virgin state. In the above example, where changes in five of the six *Q*-parameters were assumed, a ratio of 55 was demonstrated.

For the sake of demonstration we may return to the previous example ($Q \approx 0.002$) and assume that drainage and extensive pre-grouting managed to improve quality to 0.02, or even to 0.2 respectively. From Figure 44, gradient (–)*m* will perhaps have improved from (–)0.83 (when $Q = 0.002$) to (–)0.63 (when $Q = 0.02$) to (–)0.4 (when $Q = 0.2$). When corrected for tunnel size, CLI, *q*% and *n*% according to Equation 24, these two reduced gradients of (–)*m* become (–)0.67 and (–)0.43 respectively. If we assume, arbitrarily, that cutter force can (and actually needs to) be increased through these pre-treated fault zones, the following can be calculated:

1. $m = (-)0.67$, $F = 5$ tnf, $\sigma_\theta = 20$ MPa (assumed increased by grouting)

2. $m = (-)0.43$, $F = 10$ tnf, $\sigma_\theta = 25$ MPa (assumed further increased by grouting)

$$1.\ Q_{TBM} = 0.02 \times \frac{SIGMA_1}{5^{10}/20^9} \times \frac{20}{15} \times \frac{25}{20} \times \frac{20}{5} \approx 6990\ SIGMA_1$$

$$2.\ Q_{TBM} = 0.2 \times \frac{SIGMA_2}{10^{10}/20^9} \times \frac{20}{15} \times \frac{25}{20} \times \frac{20}{5} \approx 85\ SIGMA_2$$

Since the *Q*-values have been improved by the grouting we can assume that $SIGMA_1$ has increased from 0.9 MPa to about 2 MPa (a small increase in density γ has been assumed), while $SIGMA_2$ may have increased from 0.9 MPa to about 4.4 MPa. The two new Q_{TBM} values are therefore estimated to be about 14000 and 370 respectively.

According to Equation 11, PR values will now be about 0.7 and 1.5 m/hr respectively. Decimal places are rounded in each calculation since only estimation is possible. Re-evaluating Equation 16, with the same assumed fault thickness (this may even be conservative) we obtain:

1. $Q = 0.02$, $T = (5/0.7)^{1/0.33} \approx 390$ hours ≈ 16 days

2. $Q = 0.2$, $T = (5/1.5)^{1/0.57} \approx 8$ hours ≈ 1 shift

The lesson one may be tempted to draw from these examples is that a few more days of pre-injection time might help to lift fault penetration times from a matter of weeks to a shift or two. One could then also take a legitimate risk of boring faster, using higher cutter forces, so long as necessary support could be placed, for example while the grippers are re-seated.

Before leaving this chapter on this optimistic evaluation of the potential benefits of grouting, it is prudent to again relate some of the experiences of the exceptional Pinglin tunnels. The back cover of this book shows a sketch of just one of the difficult situations that can arise in caving ground, if cutter change is nevertheless imperative.

Shen et al. (1999), particularly emphasise the difficulty of efficient pre-grouting in unstable, abrasive rock when water pressures are very high, due to frequently stuck drilling rods and lost holes. The clay gouge of fault zones cannot be injected effectively, and may hydraulically fracture at high injection pressures. Tight joints that are difficult to cement grout still leak a lot of water when joint frequencies are exceptionally high as at Pinglin. The positions and directions of grout holes unfortunately may be fixed by the rear shield configuration, making grout hole spacing too wide where actually needed. To avoid flooding hazards, Shen et al. (1999), finally recommend working non-stop over weekends and national holidays. Their tough experiences at Pinglin, generously shared in this graphic article, can be a lesson to all users of TBM.

Part 3. Logging, tunnel support, probing
and design verification

CHAPTER 29

TBM Q-logging and tunnel scale effects

The attractiveness of the TBM method of excavation besides high excavation rates (when conditions are favourable) is that the reduced level of disturbance compared to drill-and-blast gives a reduced need of rock reinforcement (and an apparently higher Q-value) over a significant central range of rock qualities. When, however, the rock is of very poor quality, almost equal stability conditions (overbreak and other problems) will be experienced by the two methods of excavation and a common, lower Q-value will tend to be in operation. When the rock is of very good quality, no problems with rock support arise in either the TBM or drill-and-blast tunnel, and the Q-values can be considered the same in each case.

In the Svartisen road tunnel, a rather unique opportunity was presented to map the 6.25 m diameter TBM tunnel and later to repeat the mapping when the same tunnel was enlarged by drill-and-blast to a 7.5 m diameter road tunnel (see Fig. 48). Løset (1992) found that on average the Q-values mapped in the TBM tunnel were 1.5 to 3.0 times higher than the values mapped in the drill-and-blasted version of the same tunnel. However, this mismatch was only experienced where the TBM Q-values were in the middle range of 4 to 30 (i.e. in the drill-and-blast tunnel they were about 2 to 10). Both above and below these ranges, the Q-values were very similar with both types of tunnel excavation.

Nearly a doubling of the quantities of rock bolts and shotcrete were specified for the drill-and-blast version. Predicted permanent rock support needs for the TBM related rock conditions amounted to 1500 bolts and 1120 m^2 shotcrete (5 to 10 cm thick) for the 4400 m of marble, micaceous gneiss and meta-sandstone.

Predicted permanent rock support needs (and those finally used by the contractor) for the hybrid tunnel depicted in Figure 48 amounted to somewhat greater quantities of support: 2315 bolts and 2030 m^2 shotcrete (5 to 10 cm thick). Overburden exceeded 750 m for 2 km of the tunnel.

It appears from other experience when mapping TBM tunnels that this 'central threshold' of Q-values (where there are significantly different Q-value estimates) occurs over a lower range of rock qualities in small span tunnels, and the differences here may also be more marked. For example, drill core or drill-and-blast estimates of Q-values in the range 0.5 to 2.0 may be judged to be significantly higher (i.e. as much as $Q \approx 2$ to 10) in 2 to 3 m size pilot tunnels driven by TBM. Conversely, the 'central threshold' occurs at a higher range of Q-values in the

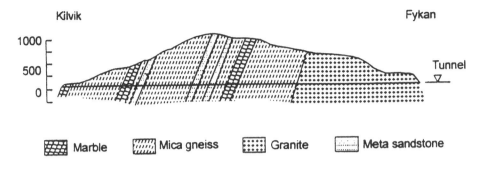

Longitudinal section with geology

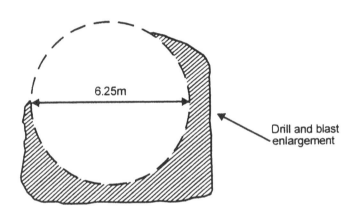

Cross section of the TBM tunnel

Figure 48. Enlargement of TBM tunnel by drill-and-blast allowed Q-values to be properly compared (Løset 1992).

case of large 8-12 m TBM tunnels. The Q-values encountered in the large diameter Channel Tunnel TBM UK drives, which were generally in the range 4 to 10 in the early kilometers of overbreaking rock, were below the 'central threshold' (where $Q_{obs.}$ becomes markedly higher than Q_{actual}) and are therefore comparable with drill-and-blast estimates.

These differences in judging Q-values are due to the maintenance of limited joint lengths in the case of TBM excavation, and the general impression of higher RQD, lower J_n and lower (better quality) J_a-values. It is difficult to spot thin clay coatings and fillings until (or unless) they are exposed by overbreak, or by swelling. Incipient joints tend to remain incipient in TBM tunnels, while they

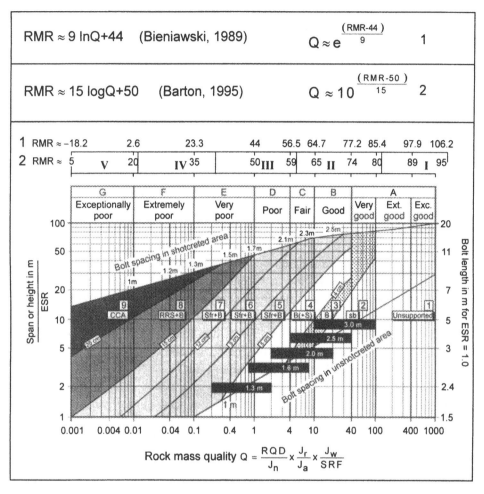

| RMR ≈ 9 lnQ+44 (Bieniawski, 1989) | $Q \approx e^{\frac{(RMR-44)}{9}}$ | 1 |

| RMR ≈ 15 logQ+50 (Barton, 1995) | $Q \approx 10^{\frac{(RMR-50)}{15}}$ | 2 |

Figure 49. Black areas show where Q-values will be over-estimated in logging of TBM tunnel walls (modified from Grimstad & Barton 1993.)

will tend to be propagated by drill-and-blast disturbance. A potential negative aspect of TBM excavation, however, is the maintenance of high tangential stresses close to (or at) the tunnel wall when blasting is not used for excavation. On occasion (high overburden, massive rock) SRF values may be judged to be higher in the case of the TBM tunnel, due to less favourable stress/strength ratios and experience of time-dependent stress-induced slabbing (Løset 1992).

The support recommendations given in Figure 49 relate, in the first instance, to drill-and-blast case records of which there are now some 1250 cases (Grimstad & Barton 1993). When this support chart is used as guidance for TBM tunnel support, it is necessary to adjust the rock mass Q-value to a higher value within the 'central threshold' range of rock qualities if TBM support quantities are to be estimated ahead of the tunnelling. Some experience of improved qualities registered in 'central threshold' ranges of qualities referred to earlier are listed below:

span 6-7 m $Q_{obs.} = 4\text{-}30$ $Q_{actual} \approx 1.5\text{-}10$
span 2-3 m $Q_{obs.} = 2\text{-}10$ $Q_{actual} \approx 0.5\text{-}2$

In Figure 49, black-coloured areas are shown for various sizes of tunnel, which are designed to show the 'central threshold' area where TBM Q-logging will give an over-estimated Q-value (×2 to 5) due to the logging problems discussed above. In each case it can be noted that the TBM range of $Q_{obs.}$ values crosses the no-support boundary, i.e. the above ranges are in the threshold of light-support, no-support. The rock mass quality judged from horizontal core or drill-and-blast excavation would require light systematic support, i.e. systematic bolting or systematic bolting plus unreinforced shotcrete. The relatively reduced need of support for the TBM tunnel is partly real, joints remain incipient, disturbance is slight, arched beams of rock remain intact but would be fractured into several pieces by blasting. The danger is in failing to recognise clay bearing surfaces where kinematic release is possible once softening has progressed due to the radial unloading caused by tunnelling. Løset et al. (1996) describe the case of a 7 m long wedge of rock falling onto the tail of a 3.3 m TBM, due to failure in recognising the reduced quality (i.e. the low J_r/J_a ratio). A perpendicular case of block fall-out is illustrated schematically in Figure 4 by Wanner (1980).

Outside the *central threshold*, both above and below, the rock behaves as equally massive or equally overbreaking and requiring support, whether the tunnel is TBM excavated or drill-and-blasted. The massive rock will show 'half-pipes' and give a 'perfect' profile almost equivalent to that of a TBM tunnel. The overbreaking rock on the other hand will give a profile that is structurally (joint-plane) controlled and will show great similarity between the TBM and drill-and-blast cases, and will require similar levels of support which can be readily selected from Figure 49 using correct values of ESR.

The above adjustments to Q-values will be needed if the TBM is designed to provide Q-values for support selection for a subsequent enlargement to a drill-and-blasted road or rail tunnel of large cross-section. An illustration of such a case from Japan is shown in Figure 50 together with three-dimensional discontinuum (3DEC) modelling of the relative deformation magnitudes in the two scales of tunnel.

The Q-values will be potentially overestimated in the TBM pilot tunnel, where deformations may generally be slight. Support needs will be minimal except in extremely poor ground. Use of the TBM pilot for pre-bolting of the central arch area of the large tunnel that follows (in this case of 20 m span) will need to consider the greatly increased deformation that will increase tension along the pre-installed bolting. More flexible reinforced plastic (RFP) or fibreglass bolts may be advisable for the pre-support in such cases.

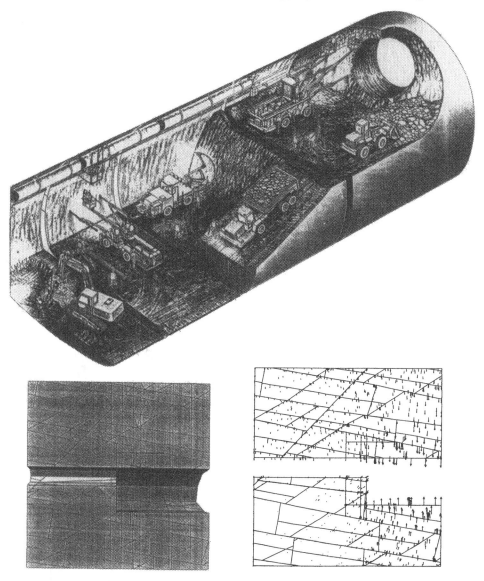

Figure 50. *Q*-values and convergence measurements obtained in a TBM pilot tunnel need careful interpretation for the prediction of full scale behaviour (Japanese Highways, and Kaynia, NGI 1995).

Scale effects are also in evidence when a TBM tunnel of small diameter (say 4 m) is used as a pilot tunnel for subsequent reaming to a larger diameter (say 12 m). The example of the Lenna Tunnel in Italy was described by Scesi & Papini (1997). Figure 51 shows some of the discontinuum (UDEC) modelling performed by these authors. When joint spacing was large (> 0.5 m) and stresses were high (> 300 m depth) and joints had shallow dip, deformations were geometrically scaled, i.e.:

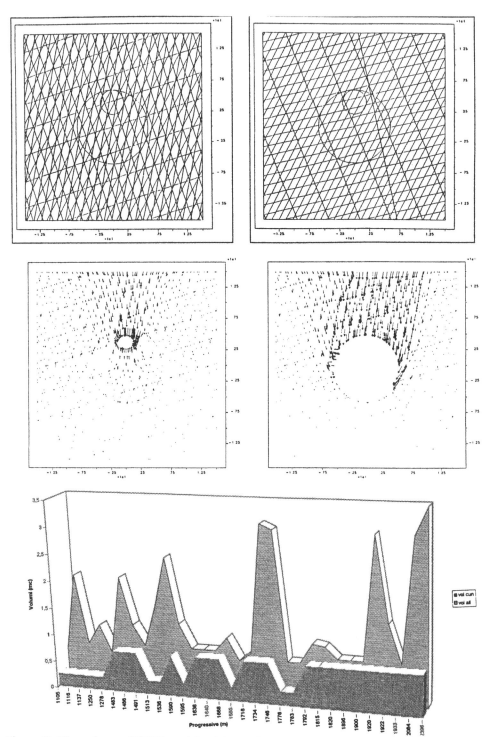

Figure 51. Discontinuum (UDEC) models of 4 m and 12 m diameter TBM tunnels, with observed differences in overbreak volumes. Lenna Tunnels (Scesi & Papini 1997).

$$\Delta_{12}/\Delta_4 \approx 3$$

However when joint spacing was reduced (0.3-0.5 m), depth was moderate (100-300 m) and joints had medium dip angles:

$$\Delta_{12}/\Delta_4 \approx 3-10$$

Scesi & Papini (1997) found that when joint spacing was small, (< 0.3 m), and depth was shallow (< 100 m) and joint dips were predominantly steep, then even greater scale effects were seen:

$$\Delta_{12}/\Delta_4 > 10$$

Figure 51 shows some contrasts in deformation fields for the two sizes of TBM tunnel. The lower diagram also shows an actual record of overbreak volumes measured in the two tunnels as they were constructed.

CHAPTER 30

Rock support methods commonly used in TBM tunnels

Tunnel support and rock reinforcement methods that were originally developed for drill-and-blast excavations, such as rock bolts, steel mesh, steel straps (chain-link-mesh) shotcrete (mesh or fibre-reinforced) and steel sets are each used in TBM tunnels. However, their application is usually delayed by the presence of the cutter head and shield. In very poor conditions, bolting and shotcrete may need to be applied ahead of the temporarily withdrawn machine such as illustrated in Figure 52.

Short shield TBM or ones with movable canopies (Fig. 3a and b) and open machines (Fig. 1) have equipment for rock bolting and steel set erection close behind the cutter head. A finger shield allows bolting or steel set erection to be carried out in a relatively protected environment, unless high inflows of water and erosion of fines is occurring, creating dangerous voids and block falls.

In such cases, greater delays are incurred and a dangerous void eventually has to be filled with concrete, as described for example by Garshol (1980). A serious case of void filling was shown by Grandori et al. (1995b), and is illustrated in Figure 37. Extreme cases of void filling in karstic limestones, even using boreholes for pumping concrete from the surface, are described by Milanovic (1997). On occasion temporary concrete bridges are needed before being able to continue TBM tunnelling at the other side of giant karstic cavities.

Without the stiffness and support provided by the void-filling concrete, tunnel stability is not usually recovered, and gripper action will be prejudiced when the same zone is reached by the grippers some time later.

In general terms, TBM tunnel support measures are applied at several specific locations from work platforms behind the cutterhead. The measures available will partly depend on the size of the tunnel. However, steel set erection and rock bolting equipment close behind the cutterhead is normally available even on small diameter TBM. Shotcreting and additional bolting facilities will usually be placed further back on one or more working platforms.

Shotcreting may need to be applied in fault-related voids ahead of the cutterhead, as described by Martin (1988), or it may be applied from a movable canopy close to the cutterhead. Otherwise it is usually not applied until more space is available further back from the face, so that rebound causes less problems to the TBM operations than it otherwise would if applied above the machine.

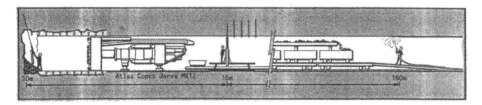

Figure 52. Support and rock reinforcement are installed at various locations in a TBM tunnel. 'Pretreatment' of a fault zone after Ilbau (Martin 1988).

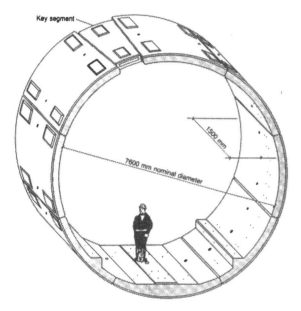

Figure 53. Concrete element expanded wedge-block lining, as used in the UK Channel tunnels of 8.4 m span (Warren et al. 1996).

When rock conditions are extremely adverse and numerous points of high pressure water inflow make shotcreting impossible, then steel arches with heavy lagging may be required. If tunnel stability is in question and a large void is in the process of forming, then continuous arches (placed flange-to-flange) or a steel plate lining as described by Garshol (1980) and Storjordet (1981) will represent two of the options that are available. Even a double-shielded machine with concrete element ring building (i.e. Fig. 53) may be stopped by such conditions, and one must revert to other 'conventional' methods, as richly illustrated in Figure 37.

Unstable ground that is suspected from probe drilling, or that is already encountered around the head of the machine, may be stabilised by spiling, by preinjection of cement or foaming grouts, by forepoling or by jet piling. Examples of top headings constructed under an umbrella of forepoling are given by Martin (1988), Grandori et al. (1995b), and Einstein & Bobet (1997), to name just three.

Excavation of the bench and newly consolidated ground may occur after several weeks (at best) or several months (more usual).

Smooth cutterheads with limited cutterhead protrusion and with closeable or partly closeable face buckets, can help to limit muck ingestion in ravelling ground such as heavily jointed, clay-bearing basalts.

Polyurethane expanding foam grouts or up-front shotcreting can also help to control the ravelling and limit void formation if water inflows are limited (or can be limited by pre-injection).

Pre-injection may be needed to prevent settlement damage caused by pore pressure reduction in overlying clays, or it may be needed to help stabilise ravelling ground. The stabilising effects of pre-grouting and its potential for increasing Q-values, shear strength, deformation modulus and seismic velocity, have been discussed by Barton (1999). Some of these aspects were discussed in Chapter 28.

The delaying effects of essential pre-injection have been well documented. Figure 36 showed the direct influence of pre-injection on advance rates at part of the 33 km of bored tunnel needed for the Oslo-VEAS sewage tunnel, as reported by Garshol (1983). A total of 387 km of pre-injection grout holes were required at this project, and 4430 tons of cement grout and 1880 m^3 of chemical grout were used (Asting 1981). In some sensitive urban areas, pre-injection cost three times more than the average tunnelling costs. These same tunnels in predominantly limestones and shales with numerous eruptive dykes required bolting, shotcreting or cast concrete lining in only 7% of their total length. This was a case of high penetration rates and low advance rates. [The open circles shown in Figure 34 illustrated the adverse effects of pre-injection on utilisation and on gradient $(-)m$.]

When erecting steel sets under a finger shield it is clear that the machine head cannot later be withdrawn unless the steel sets are temporarily removed. It may therefore be necessary to apply shotcrete and rock bolts in front of the cutterhead when very poor ground is encountered. If prior knowledge of this poor ground is gained by systematic probe drilling, then over-boring as illustrated in Figure 39 will be a solution. It will also give the necessary space for local heavy support such as cast concrete, in case most of the tunnel can be left unlined, and has not therefore been over-dimensioned for subsequent lining.

In highly automated, high advance rate, soft rock tunnels, the provision of a standard and often conservative pre-cast lining may be attractive from the point of view of schedule and future tunnel use. The expanded wedge-block lining illustrated in Figure 53 performs well if the bored profile remains more or less circular due to limited overbreak. A hexagonal element liner is reportedly becoming an attractive competitor to the conventional element shape.

When tunnel stability is in doubt and stand-up time is limited, a bolted concrete element liner may be preferable to wedge-block solutions. This can be assembled in the tail of a double-shielded machine.

Clearly these solutions to permanent lining are rather expensive in comparison with a tunnel that requires only sporadic areas of rock reinforcement and occasional heavy lining of fault zones. However, in relation to final cast concrete linings they may have certain advantages, including 'immediate' permanent support of the tunnel.

Tunnels that will require a smooth finish for water or sewage transport, or for hydropower, will often need a final concrete lining if many areas of poor ground are encountered, and over-boring has not been performed. Due to the delay involved in starting the concrete lining operation, one is usually depending upon the temporary support for a period of at least 6 months and up to several years in the case of longer tunnels or very poor rock conditions. This should be born in mind when selecting temporary support.

Once the concrete lining operation begins, very fast rates are achieved, often in the range of 1 to 2 km per month with suitable telescopic shutters. Grandori et al. (1995b), reported daily rates of concrete lining construction of 50 to 100 m, with best days of 140 m and a best month of 2120 m at the Evino-Mornos water tunnel. The four TBM used at this project provided two ready lined tunnel sections (double shield TBM with hexagonal segments) totalling some 17 km, while the two open TBM sections required the final cast concrete liner for the water transport, totalling some 12 km. Lining operations took only 5 to 6 months, despite some stoppages for extra support in areas of squeezing ground.

Since TBM tunnels have a multitude of purposes, for example sewage, water supply, hydropower, temporary pilot tunnel, road or high-speed rail, it is clear that a range of safety requirements must exist just as in the case of drill-and-blasted excavations. The ESR concept used in the Q-system for modifying the effective tunnel dimension when selecting support can also be used for support design in TBM tunnels. As shown in Figure 49, this ESR factor will modify the support needs in the direction of lighter support for temporary support of pilot tunnels. The following ESR values can be utilised in conjunction with the support recommendations given in Figure 49 (Table 18).

Table 18. Suggested ESR values for TBM support/liner selection.

Tunnel type	ESR-value
All support of temporary nature	1.5 × ESR and 5Q*
Pilot tunnels	2.0
Water/sewage tunnels	1.5
Traffic tunnels	0.5 to 1.0**

*Modify factor 5Q to 2.5Q if temporary support must stand for more than one year. **ESR may be reduced to 0.5 for long, high speed rail or long motorway tunnels. Note: Q-value correction (×2 to 5) is needed in the 'central threshold' areas for the relevant tunnel diameters (Fig. 49). Use 2Q in large diameter tunnels and 5Q in small diameter tunnels to account for the greater stability of TBM tunnels in this region of the Q-system support chart.

CHAPTER 31

Some support design details for TBM tunnels

The potential use of direct Q-system support recommendations for TBM tunnels has been discussed in Chapters 29 and 30. The inherently greater stability of circular, machine-excavated tunnels, and their reduced need for support is taken care of by the 'central threshold' Q-factoring (×2 to 5) discussed earlier and shown in Figure 49. Outside this zone, to the right-hand side (higher Q-values), no temporary or permanent support will be required. To the left-hand side (lower Q-values), equally heavy temporary and permanent support will be required in the TBM tunnel as in the drill-and-blasted tunnel.

The extensive case record basis behind Figure 49 (1250 cases: Grimstad & Barton 1993, Barton & Grimstad 1994) suggests that there will be great benefit from utilising the Q-system recommendation together with the RMR-related stand-up time techniques described in Chapter 19 and in Figure 42. However, some added details are required in view of the different locations where support can (or cannot) be provided in the more confined space of the TBM driven tunnel.

Data for some Italian TBM tunnels reviewed by Scolari (1995) emphasise the reduced utilisation of the TBM due to time spent in applying tunnel support, break of routine and so on. Figure 54 shows the dramatic reduction in daily TBM utilisation, as number of rock bolts per meter increases with reducing rock class. Even when using temporary Swellex bolts, which are easy to install, delays may be incurred.

The Austrian-based rock classes F1 to F6 shown in Figure 55 have been 'fitted' with RMR and Q-values in interesting review articles by Martin (1988) and Scolari (1995). This approach will be followed here also, as the Austrian classification (Ö-Norm B2203, modified by Ilbau: Scolari 1995) and support selection method has many useful features.

Since daily advance rates and utilisation are each given for common time periods in Figure 54, we can utilise the following equations to back-calculate the penetration rates and Q_{TBM} values for the six Austrian classes.

$$AR = U \times PR \qquad (1)$$

$$U = T^m \qquad (9)$$

$$AR \approx 5Q_{TBM}^{-1/5} \times T^m \qquad (12)$$

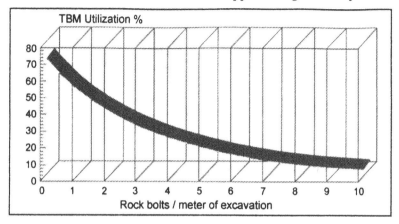

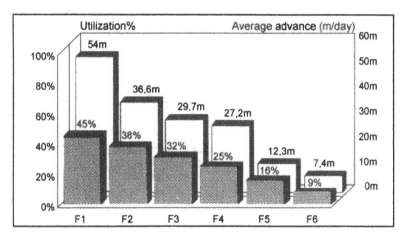

Figure 54. TBM utilisation as a function of bolting needs and rock class, at the 12.5 km Prato Isarco pilot tunnel in Italy (Scolari 1995).

From these earlier equations we can derive the following:

$$Q_{TBM} = (5/PR)^5 \qquad (25)$$

Since $U = T^m$, we can also write $m = \log U/\log T$ and use this to back-calculate the gradients of deceleration caused by different support methods.

The increasing negative gradients $(-)m$ are in general consistent with the Q-m trend shown in Figure 45, and provide a quite realistic range of Q-values in the lower rock classes. Gradient $(-)m$ as high as $(-)0.25$ and $(-)0.30$ in classes F1 and F2 suggest that the small size of the pilot tunnel ($\varnothing = 3.5$ m) may have negatively influenced efficiencies. The gradients given in Table 19 have been plotted in Figure 35 over the first time-range of 1 hr to 24 hrs. They provide another source of Q-value estimation in the area 'unexpected events'.

In view of the fact that rock support measures may have such a strong influence

Class	Approx. Q range	Approx. RMR range	Typical Section Diameter 6m	Rock Mass Behaviour
F1	10-100	65-80	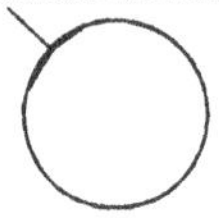	Long term stability
F2	4-10	59-65	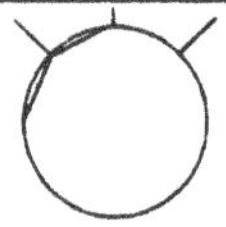	Local rockfall
F3	1-4	50-59	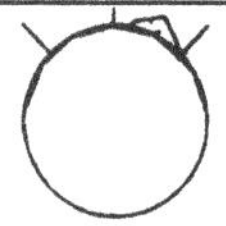	Frequent rockfall in machine area
F4	0.1-1	35-50	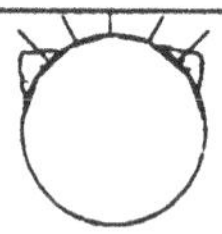	Frequent rockfalls in machine area
F5	0.03-0.1	27-35	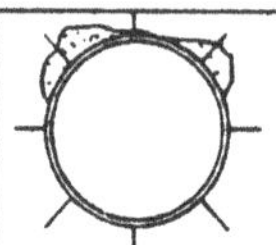	Frequent rockfalls in cutterhead area after each stroke
F6	0.01-0.03	20-27	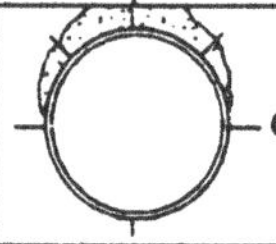	Large overbreak in cutterhead area after partial strokes
F7	0.001-0.01	5-20	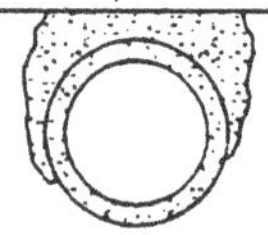	No self supporting capacity

Figure 55. a) Austrian (Ilbau-modified) TBM support scheme, with writer additions of Q and RMR ranges in relation to classes F1 to F7 (after Scolari 1995).

Table 19. Back-analysis of Prato Isarco pilot tunnel data (Fig. 54).

Average m/day	54	36.6	29.7	27.2	12.3	7.4
Average AR m/hr	2.25	1.53	1.24	1.13	0.51	0.31
Utilisation	0.45	0.38	0.32	0.25	0.16	0.09
Class	F1	F2	F3	F4	F5	F6
PR m/hr (Eqn 1)*	5.0	4.0	3.1	4.5	3.2	3.4
$m = \log U/\log T$	(–)0.25	(–)0.30	(–)0.36	(–)0.44	(–)0.58	(–)0.78
Q_{TBM} (Eqn 25)*	1.0	3.1	10.9	1.7	9.3	6.9
Approx. Q from Fig. 45	0.55	0.32	0.22	0.13	0.04	0.005

*These PR values and calculated Q_{TBM} values show obvious operator influence, with PR showing little relation to rock class in this case record.

Class	SUPPORT MEASURE			Influence on advance
	Type	Quantity per linear metre	Place of installation	
F1	Local Support Rockbolts L=2.0 m required	Up to 0.5	Working platform	None
F2	Local Support Rockbolts L=2.0 m Wire mesh Shotcrete 5 cm	Up to 1 Up to 1.0 m² Up to 0.1 m³	Working platform	None
F3	System Support Rockbolts L=2.0 m Wire mesh Shotcrete 5 cm	From 1 to 3 From 1 to 1.5 m² From 0.1 to 0.5 m³	Working platform	Short delays
F4	Rockbolts L = 2.5 m Wire mesh Shotcrete 8 cm Steel ribs	From 3 to 5 From 5 to 9 m² From 0.5 to 1.0 m³ From 40 to 80 kg	Working platform behind cutterhead	Delays after each stroke
F5	Rockbolts L = 2.5 m Wire mesh Shotcrete 10 cm Steel ribs	From 5 to 7 From 9 to 18 m² From 1.0 to 1.8 m³ From 80 to 160 kg	Immediately behind cutterhead after each stroke, additional support from working platform	Long delays after each stroke
F6	Rockbolts L = 3.0 m Wire mesh Shotcrete 15 cm Steel ribs	From 7 to 10 From 18 to 27 m² From 1.8 to 3.0 m³ From 160 to 300 kg	Immediately behind cutterhead after each partial stroke, additional support from working platform	Long delays after each partial stroke
F7	Special measures to be decided according to conditions	e.g, spiling, pre-injection, forepoling, jet grouting, foam injection, cast concrete		Delays of months or more

Figure 55. Continued. b) Support types for classes F1 to F7 (after Scolari 1995).

on utilisation and advance rate, and that there are limitations to *what* can be applied *where*, it is useful to split the rock support measures, and indicate where each should be applied. This has been successfully achieved in the Austrian Ö-Norm method, which has been discussed in the English language literature by Martin (1988) and Scolari (1995) among others. Both these authors presented the Austrian F1 to F7 classes of rock support, and showed some approximate relations to Q-values and RMR values.

The scheme shown in Figure 55 is mostly based on an Ilbau modification of the Ö-Norm B2203 method, as presented by Scolari (1995). Symbolic sketches of overbreak have been added, and what is considered a more logical range of Q and RMR values are also given here. This is based on cross-referencing with the Q-system (Fig. 49) support methods for an approximately 6 m diameter tunnel, which can be read from Figure 49. The inter-relationships between Q and RMR are based on the second equation listed at the top of the same figure, from Barton 1995.

Modifications to the support-quantities such as bolt spacing and shotcrete thickness for different tunnel sizes and different tunnel uses can be made by integrating Figures 49 and 55 with Table 18.

Although steel mesh reinforcement is a valuable temporary measure for bridging between bolts or steel sets in the usually conveniently cylindrical shape of a TBM tunnel, it is urged that fibre-reinforced shotcrete S(fr) be used where steel reinforced shotcrete is needed. Recently developed alkali-free accelerators allow unlimited build-up of S(fr) which may be equally beneficial ahead of the cutterhead in crisis situations, and as permanent lining applied behind the back-up rig. If steel mesh has been applied behind the finger shield to bridge between bolts, then it must first be covered with plain shotcrete to minimise rebound.

It will be noted in Figure 55 that a 'place of installation' column is given on the right-hand side of the figure. In support classes F1 to F3, one of the work platforms is used for bolt, mesh and shotcrete application, while in poorer conditions support application has to be right behind the cutterhead due to limited stand-up time for the F4 to F6 classes. When class F7 is encountered, various pre-injection and pre-reinforcement methods will be used together with eventual hand excavation of collapsed or eroded material from within and ahead of the machine. Very serious delays are an almost inevitable result (i.e. $m \approx -0.7$ to -0.9) when Q-values are lower than 0.01. The gradients suggest utilisation of less than 1% in such conditions, if the delay exceeds weeks and even months.

CHAPTER 32

Probing and convergence measurement

The big investment in a sophisticated TBM and the expectation of mostly rapid advance rates can each be spoiled by the unexpected delays caused by 'unexpected ground'. Only a few percent of the total length of a tunnel may be 'unexpected', yet these few percent could double the construction time in some cases. Although deep tunnels are notoriously difficult to predict, regular probe drilling during cutter change and maintenance shifts could largely remove the unexpected; especially if performed with two slightly diverging probe holes.

Figure 56 shows how probe hole logging could be used. Measurement while drilling and eventual sonic logging, or drill core and Q-logging would be supplemented by side-wall logging behind the advancing cutterhead, to confirm or correct the information acquired perhaps one day before and 25 m or more ahead of the present face position. The preliminary Q-value class (Figs 49 and 55) will determine the first rock reinforcement to be installed immediately behind the cutterhead, at (A) or eventually later at (B) or (C). In the example shown, the class 4 permanent support for this hard rock tunnel of 8 m diameter was *B 1.5 c/c plus S(fr) 12 cm* (see key to symbols at the bottom of Fig. 56). The component 'B 1.5' is applied at work platform (A) (where steel sets could also be erected). Shotcreting is performed first at work platform (B) (4 cm) then completed at (C) (8 cm more).

In this example, an expectation of deformation of 20 to 50 mm based on the assumed Q-value range of class 4, has been more or less confirmed by convergence monitoring. In a large open TBM tunnel there is space for this monitoring to be done routinely. As noted towards the bottom of Figure 56, if convergence (Δ) is more or less than expected between (A) and (B) and between (B) and (C), then respectively more or less support/reinforcement is added at (B) and (C) to compensate for the poorer or better behaviour. In this tunnel, because of its critical nature, we have combined the predictive abilities of the Q-system logging (the NMT philosophy) with the observational approach common in NATM. Barton & Grimstad (1994) discussed the differences between NMT and NATM in some detail. Here we have combined the merits of both approaches. This method is consistent with the philosophy behind the Austrian method shown in Figure 55, but was developed independently from this Ö-Norm classification scheme.

Although the use of monitoring data can be criticised as being idealised, and a

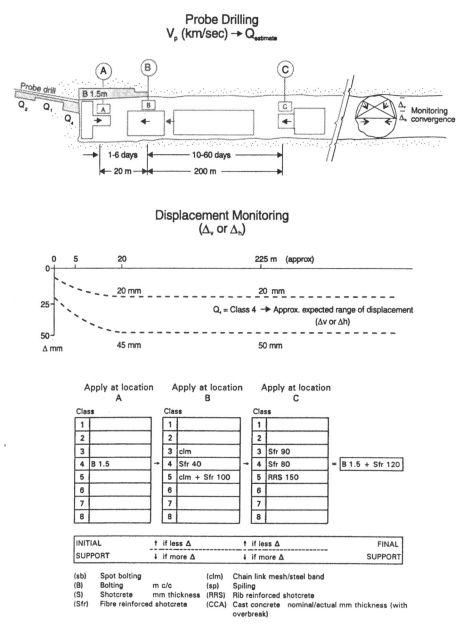

Figure 56. Preliminary classification by probe drilling, convergence monitoring and Q-logging, and application of optimal support measures at work platforms A, B and C in a large diameter TBM project (Barton & Grimstad 1994).

stable deformation-time record is actually no guarantee of stability, it is nevertheless one of the tools at our disposal besides probe drilling (e.g. Fig. 7) and sonic logging. Figure 57 illustrates an idealised deformation-time curve. In practice it may be difficult to start monitoring before working platform (A). In this case only the shaded (cross-hatched) portion of the Δ-T curve (or Δ-length curve) will be registered. In principle, different magnitudes of deformation will signify different values of Q as indicated in the lower half of Figure 57.

The reason that Q and Δ are linked is shown in Figure 58. The Q/SPAN-deformation data has been assembled from case records since 1980, and was up-

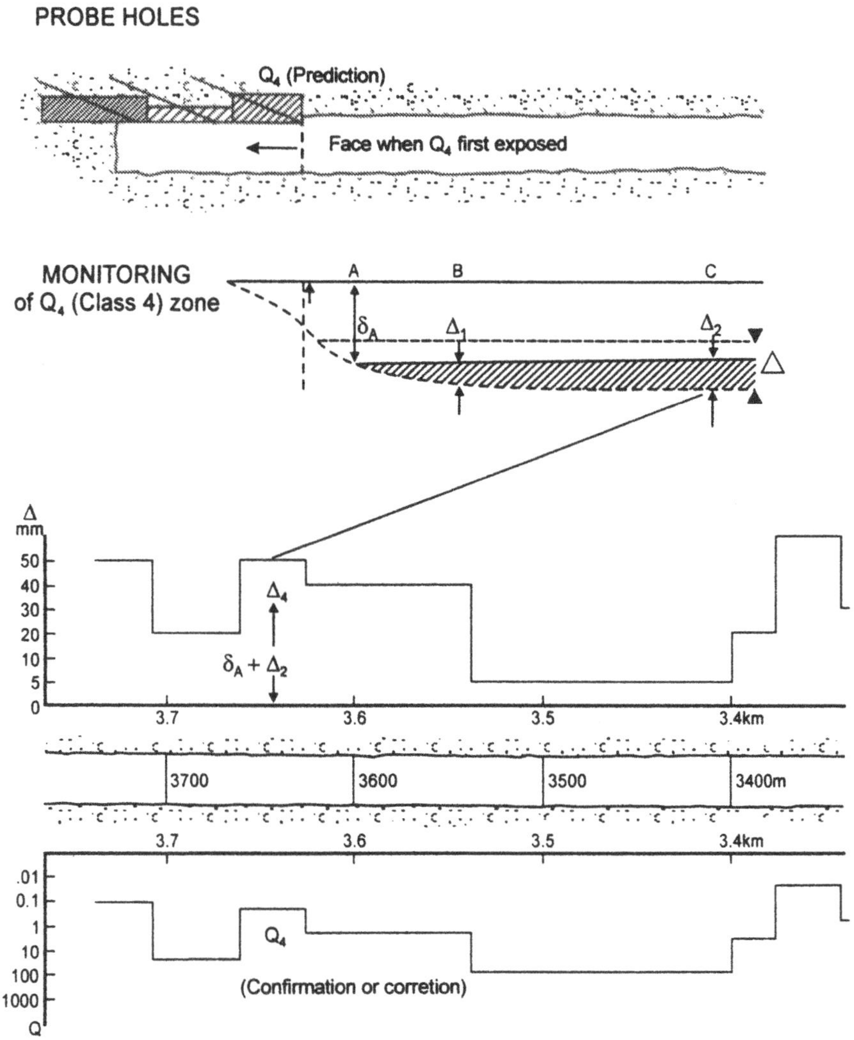

PROBE HOLES

Figure 57. Idealised relation between Q-value and deformation, for confirmation or correction of probe-hole classification.

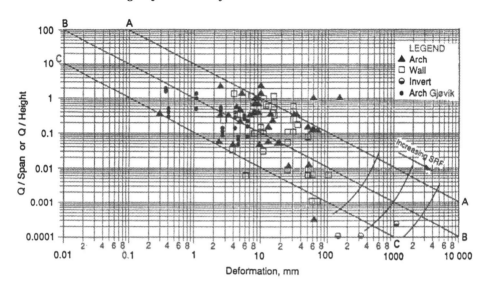

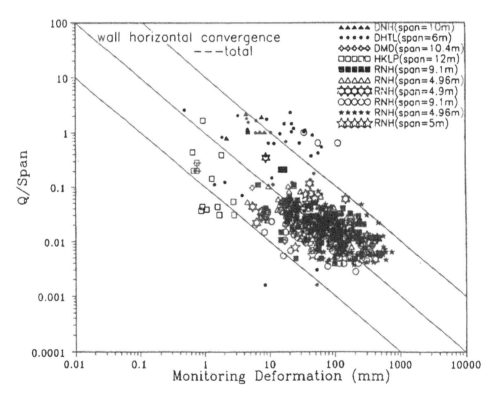

Figure 58. Q/Span or Q/diameter versus deformation (Barton et al. 1994, Chen & Kuo 1997).

dated in 1994, and greatly supplemented by soft and faulted rock data from Chen & Kuo (1997) from Taiwanese tunnel projects. The central trend of behaviour is classically simple in form:

$$\Delta \text{ mm} \approx \frac{\text{SPAN(m)}}{Q} \tag{29}$$

Some of the scatter can probably be explained by the modified form suggested by Barton (1998) in which the vertical and horizontal deformations are related to tunnel span and height respectively, and to the ratio of vertical and horizontal stress in relation to the uniaxial strength of the rock.

$$\Delta_v \approx \frac{\text{SPAN}}{100Q} \sqrt{\frac{\sigma_v}{\sigma_c}} \tag{30}$$

$$\Delta_h \approx \frac{\text{HEIGHT}}{100Q} \sqrt{\frac{\sigma_h}{\sigma_c}} \tag{31}$$

where Δ_v, Δ_h, SPAN and HEIGHT are in millimeters.

For a circular TBM tunnel of diameter D (mm), the deformations will be anisotropic if the horizontal to vertical stress ratio K_o is $\neq 1.0$.

$$\Delta_v \approx \frac{D}{100Q} \sqrt{\frac{\sigma_v}{\sigma_c}} \tag{32}$$

$$\Delta_h \approx \frac{D}{100Q} \sqrt{\frac{\sigma_h}{\sigma_c}} \tag{33}$$

$$K_o = \frac{\sigma_v}{\sigma_h} \approx \left(\frac{\Delta_h}{\Delta_v}\right)^2 \tag{34}$$

The tunnel and cavern monitoring data collected in Figure 58 usually has the same limitations as the shaded (cross-hatched) area in Figure 57. The absolute deformation cannot be captured without using pre-placed borehole extensometers or inclinometers. Probably the absolute deformation could be closely related to Q_c, using the modified approximation for rock mass strength (modified from Singh 1993).

$$\text{SIGMA}_{cm} = 5\gamma Q_c^{1/3}$$

We may therefore expect a range of absolute displacements roughly in accordance with the following:

$$\delta_v(\text{absolute}) \approx \frac{D}{100Q}\sqrt{\frac{\sigma_v}{\text{SIGMA}_{cm}}} \tag{35}$$

$$\delta_h(\text{absolute}) \approx \frac{D}{100Q}\sqrt{\frac{\sigma_h}{\text{SIGMA}_{cm}}} \tag{36}$$

Example
$D = 8$ m, depth $= 100$ m, $\sigma_v \approx 2.5$ MPa, $\gamma = 2.5$ tnf/m^3, $Q = 1$, $\sigma_c = 50$ Mpa.
From Equation 32:

$$\Delta_v \approx \frac{8000}{100 \times 1}\sqrt{\frac{2.5}{50}} \approx 18 \text{ mm}$$

From Equation 7:

$$\text{SIGMA}_{cm} \approx 5 \times 2.5 \times \left(1 \times \frac{50}{100}\right)^{1/3} \approx 10 \text{ MPa}$$

From Equation 35:

$$\delta_v(\text{absolute}) \approx \frac{8000}{100 \times 1}\sqrt{\frac{2.5}{10}} \approx 40 \text{ mm}$$

On the face of it, such deformations appear very reasonable, giving an expectation in this case of a realistic 45% to 50% residual deformation from the cutterhead monitoring position, where about half of the absolute elastic deformation has already occurred.

When using Equations 29 to 36 with extreme Q-values (i.e. 0.01 or 0.001), extremely high deformations will be predicted, perhaps signifying tunnel closure, unless necessary spiling, pre-injection and other measures actually artificially increase the Q-value. Conventional support such as steel sets or shotcrete do not *increase* the Q-value, but perhaps prevent it from reducing further. The improved, pre-treated Q-values will therefore tend to be most relevant, as discussed earlier in another context (Chapter 28).

CHAPTER 33

Probing and seismic or sonic logging

As indicated in the foregoing chapter and in Figure 56, a systematic approach to probe drilling during cutter change and maintenance shifts would also allow a systematic approach to sonic logging of the hole, and 'VSP'-style tunnel-to-probe hole velocity measurements.

Various forms of sonic and seismic logging performed at Japanese tunnels are described by Mitani et al. (1987) (see PR-V_p relation in Fig. 30), Mitani (1998), and Morimoto & Hori (1986), who logged the velocities in the disturbed zone around a small diameter (2.6 m) TBM headrace tunnel for a hydroelectric project. The loosened zone of some 20 to 60 cm thickness showed velocities as low as 0.66 to 1.0 km/s, compared to undisturbed velocities of 2.6 to 3.5 km/s. Besides radial stress relief, drying out of the joints in the EDZ might have been a factor here (e.g. Barton 2000).

The specific purpose of sonic logging ahead of the tunnel is to obtain data ahead of time as discussed by Barton (1996) (see Fig. 59), and as shown in a very interesting example by Nishioka & Aoki (1998) (see Fig. 60).

Interpretation of the rock quality from the seismic velocity obtained from confined (stressed) rock masses ahead of the tunnel needs careful consideration. Barton (1991, 1995), developed a simple relationship between the Q-value and the shallow refraction seismic P-wave velocity (for hard rocks of low porosity) as follows (this has been referred to in some earlier chapters):

$$V_p \approx 3.5 + \log Q \tag{37}$$

This was subsequently modified for a wider range of rock types and stress levels in the format shown in Figure 61. To the first two basic equations shown in the top of the figure, we may add the subscript (c) where $Q_c = Q \times \sigma_c / 100$ (i.e. normalised by uniaxial strength) as previously.

The heavy black line and circles shown in the figure represent the shallow, hard rock relation described by Equation 37. Corrections to V_p (negative adjustments) are made for porosities larger than 1% and positive adjustments are made for depths (or equivalent stresses) greater than 25 m. The strongly non-linear stress effect on V_p and the nearly linear effect of porosity are consistent with numerous data reported in the literature (Barton 2000).

In the example shown in Figure 61, from the Swedish Hard Rock Laboratory at

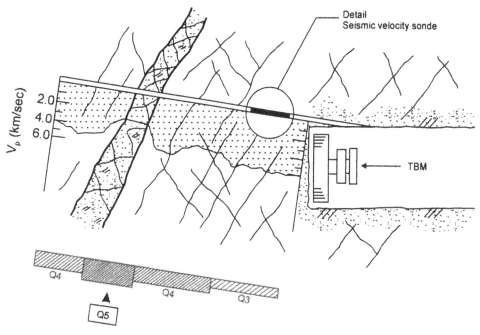

Figure 59. Probe drilling and seismic velocity logging for preliminary rock class estimation (Barton 1996).

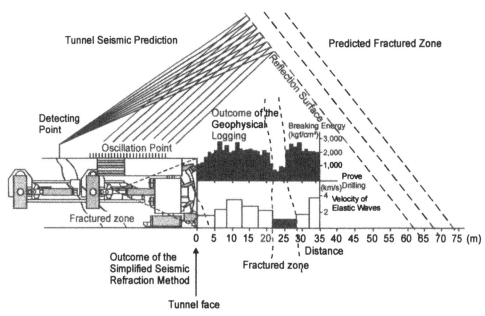

Figure 60. Seismic refraction and reflection measurements, and sonic logging method proposed by Kajima Corp., Japan (Nishioka & Aoki 1998).

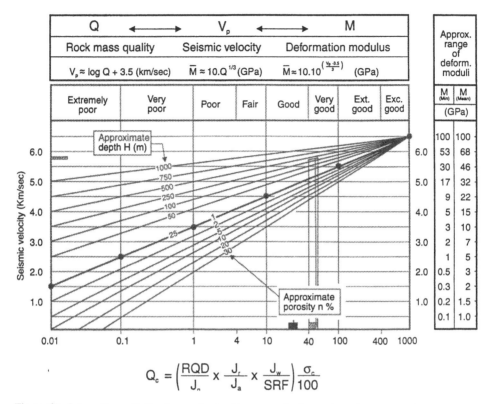

$$Q_c = \left(\frac{RQD}{J_n} \times \frac{J_r}{J_a} \times \frac{J_w}{SRF} \right) \frac{\sigma_c}{100}$$

Figue 61. Integration of Q-value, seismic P-wave velocity and rock mass static deformation modulus (Barton 1995, 1999).

Äspö, the mean Q-values for the diorites and granites based on NGI logging of 830 m of core were approximately 20 and 25. Correction for $\sigma_c \approx 200$ MPa gives $Q_c \approx 40$ to 50, as shown by cross-hatching in the same figure. Cross hole seismic measurements in the walls of the 420 m depth tunnels (TBM and drill-and-blast) gave seismic velocities generally in the range 5.8 to 6.1 km/s, and measured convergences corresponded to deformation moduli of about 60 GPa in this rather good quality rock.

The above example is given to illustrate the potential pitfall of assuming typical shallow seismic refraction velocities and their usual relation to Q-values (Equation 37). In the above example a much higher Q-value than reality might erroneously have been assumed, based on the measured V_p which was as high as 6 km/s.

The discrepancy caused by depth (or stress) and porosity increases as the rock mass quality reduces, due to the joint closure and general compaction effects of the high stress if the rock mass is not yet disturbed by the approaching tunnel face.

This also emphasises the desirability of sonic logging *at least* a diameter ahead of the face as a minimum, preferably more than a day's advance in front (i.e. 25 to

50 m in poorer rock) so that there is adequate time to react to the correct information. A radically reduced V_p and Q-value and a decision for pre-injection, pre-reinforcement and perhaps overboring may save weeks or months of delay and cost only one or two days in 'lost' production.

CHAPTER 34

Verifying TBM support classes with numerical models

Unfortunately in most TBM projects where stability problems are encountered and large delays occur, the necessity for modified support techniques is discovered after the problem has arisen. The same is true for TBM design changes, such as extended finger shields (length and angular coverage). This situation may arise due to inadequate investigations or due to over-optimistic consultants, owners and contractors who are each pressured by time constraints.

In a few cases, back-analyses are performed while the TBM is at a stand-still, and while the owner is planning a new contract for a new contractor with a differently designed TBM!

The *only* 'advantage' of this 'wait-and-see' approach is that the numerical back-analysis can be more accurate, due to better information on the geology, joint properties and apparent stress levels. Improved investigations and visualisation of potential problems with simplified numerical models may help all parties to avoid some of the potential problems, when there are doubts about certain rock classes and support designs. In effect, the modelling is done to verify the support designs that may have been derived from Figures 49 and 55 or from alternative design methods.

The modes of rock mass behaviour that cause most problems in tunnelling, especially TBM tunnelling, will seldom be captured by continuum models even when anisotropic stresses are assumed. Obvious exceptions to this claim would be highly stressed hard massive rocks that are expected to show stress-slabbing and perhaps bursting, and very soft altered or sheared materials, or weak sedimentary rocks with limited jointing and widely spaced bedding that are expected to squeeze.

Much more common stability problems will be caused by the discontinuous and anisotropic nature of most rock masses, by too many joint sets, by clay coatings, faulting and erosive water inflows.

It is not the intention here to describe numerical methods and the input data required in any detail. Instead, attention will be focussed on one method (UDEC, Cundall & Hart 1993) and on one constitutive law for the rock joint behaviour (Barton & Bandis 1990). Methods of estimating the necessary strength and deformation properties of the joints are based on simple index tests that can be conducted on joints recovered by core drilling. Corrections are made for the full-scale

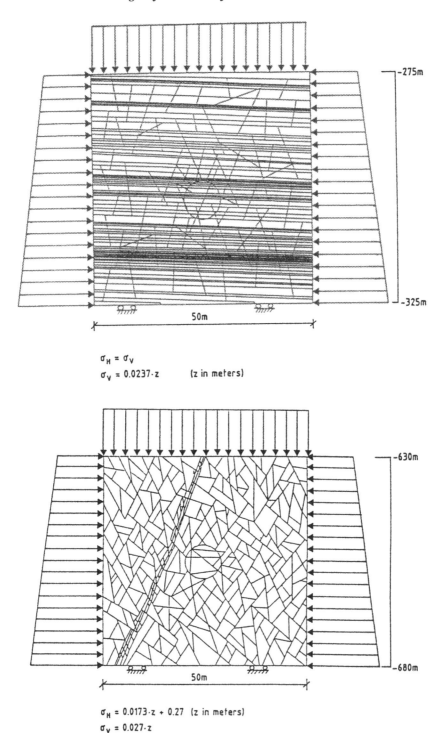

Figure 62. Boundary stress and jointing assumptions for two-dimensional UDEC-BB models of an 8 m diameter TBM tunnel in sandstones and tuff-ignimbrite (NGI 1991).

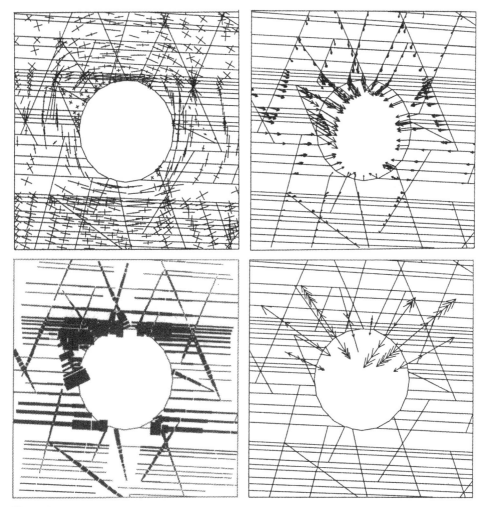

Figure 63. Stresses (max. 55.1 MPa), displacements (max. 15.0 mm), joint shearing (max. 8.0 mm) and bolt loadings (max. 15.0 tnf) in an 8 m diameter TBM model using UDEC-BB (Chryssanthakis, NGI 1991).

rock mass with blocks of different dimensions. The numerical and constitutive models have been combined within the Cundall-Itasca-NGI code and sub-routine called UDEC-BB, which has been in regular use in several countries since 1985. In 1995, a very realistic fibre-reinforced shotcrete code was added by Itasca which we have termed UDEC-S(fr).

Figure 62 illustrates the boundary stress assumptions and joint geometries that were assumed for an 8 m TBM tunnel that was planned as an access tunnel to a possible nuclear waste repository in England (UK Nirex Ltd). The upper model, with the tunnel centred at about 300 m depth in a sparsely jointed, bedded sandstone contrasts with the lower model, where the tunnel is at about 650 m depth in

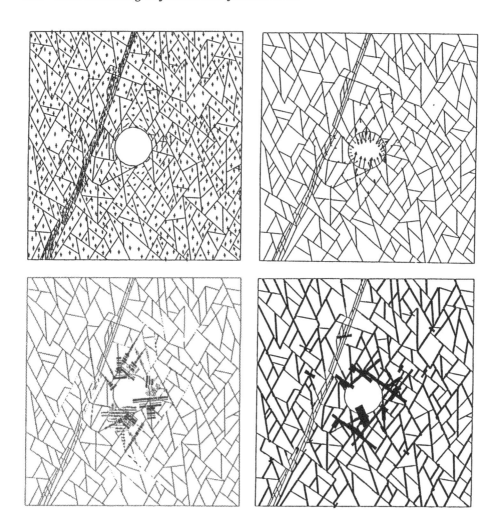

Figure 64. Stresses (max. 30.1 MPa), displacements (max. 10.6 mm), joint shearing (max. 7.6 mm) and hydraulic apertures (max. 2.0 mm) in an 8 m diameter TBM model using UDEC-BB (Hansteen, NGI 1991).

jointed and sometimes faulted tuff and ignimbrite. Properties for each of the joint sets were obtained from simple tilt tests (self-weight shearing) and Schmidt hammer tests on joint samples recovered by deep drilling. Surface mapping of relevant outcrops helped to define the idealised joint patterns.

Following consolidation, numerical excavation and numerical installation of 8 rock bolts after some 50% of the final deformation had occurred, the graphic output shown in Figures 63 and 64 was obtained. Principal stresses, deformation vectors, joint shearing and hydraulic apertures are shown that clearly indicate some anisotropic loading of the tunnel perimeter.

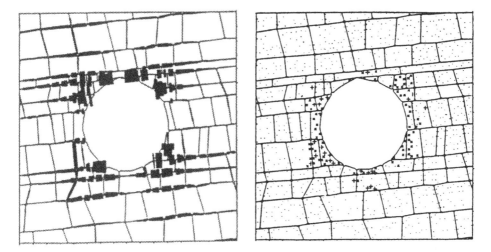

Figure 65. Excessive joint shearing and yielding of intact rock at a TBM tunnel in shale, modelled by UDEC-BB (Bandis 1990).

In the case of the bedded sandstone model, occasional wedge formation in the walls, and general horizontal convergence are seen, despite the isotropic ($\sigma_h = \sigma_v$) stress distribution. Bedding plane shearing would have been greatly accentuated if weak, clay coated beds had been modelled.

In the case of the jointed tuff/ignimbrite model, vertical convergence domi-nated with $\sigma_v > \sigma_h$ (as shown) while another section of the same spiralling tunnel with $\sigma_h > \sigma_v$ (due to horizontal stress anisotropy) showed horizontal convergence as the dominant mode. The excavation disturbed zone (EDZ) caused by stress re-distribution, block displacement, joint shearing and joint aperture change are most of the reasons for reduced seismic velocities and for reduced deformation moduli that are unknowingly or knowingly registered by the grippers some 10 or 15 m behind the face.

A third UDEC-BB model of a TBM tunnel in squeezing, yielding shale that was reported by Bandis (1990), is shown in Figure 65. Shearing along the weak bedding planes and over-stressing of the relatively weak shale around the highly stressed periphery was combined, in practice, with accentuated invert heave caused by swelling. The considerable difficulties encountered by the contractor could probably have been largely predicted if similar analyses had been requisi-tioned ahead of time.

CHAPTER 35

Logging rock quality and support needs

The degree of thoroughness with which TBM tunnel progress is logged in relation to rock conditions depends on many potential factors. The presence of continuous steel sets or pre-cast linings obviously forces geotechnical logging to be performed at the face (following partial withdrawal of the cutterhead) and along a narrow band of sidewall at the face and immediately behind the cutterhead 'through' the finger shield. Examples of such logging and a good illustration of the confined conditions were given by Warren et al. (1996), from the UK Channel tunnel, which was driven during the early kilometers through jointed, overbreaking chalk marl, with poor to fair Q-values and difficult ring building conditions.

In a large diameter open machine, logging conditions are more favourable, and the inherent absence of a continuous liner (at least while tunnel driving) means that logging can be performed at many locations from the cutterhead area and backwards, with good access to the arch and to the lower walls above the frequently placed concrete elements in the invert.

A scheme for logging rock conditions, pre-treatment, leakage and support needs (or support already applied) that can be used in a large, open TBM tunnel is illustrated in Figure 66, where stations A, B and C are shown. Station A' implies the face itself or even ahead of the face if overbreak/ravelling is occurring, and if 'pre-treatment' is applied here to try to control a potentially worsening situation.

It will be noted that there are many spaces on the chart for signing off the recorded conditions, together with the date. This presupposes an atmosphere of cooperation between the Contractor's engineering geologist and the Owner's engineering geologists. Each must have previous experience from tunnel logging and TBM work, and have experience with the logging method chosen for the project.

If conditions are logged alone by the separate parties, perhaps using different techniques (e.g. Q and RMR separately), there is unlikely to be agreement, nor the climate for cooperation when real difficulties arise (i.e. $Q \leq 0.01$). The natural tendency for 'pulling in opposite directions' (the rock is 'better or sufficiently jointed' according to the Owner, the rock is 'worse or insufficiently jointed' according to the Contractor) is effectively neutralised when both parties log and sign off together.

Logging in TBM tunnels, even behind the back-up rig with good lighting, is not as easy as in a drill-and-blasted tunnel, due to the frequent lack of significant

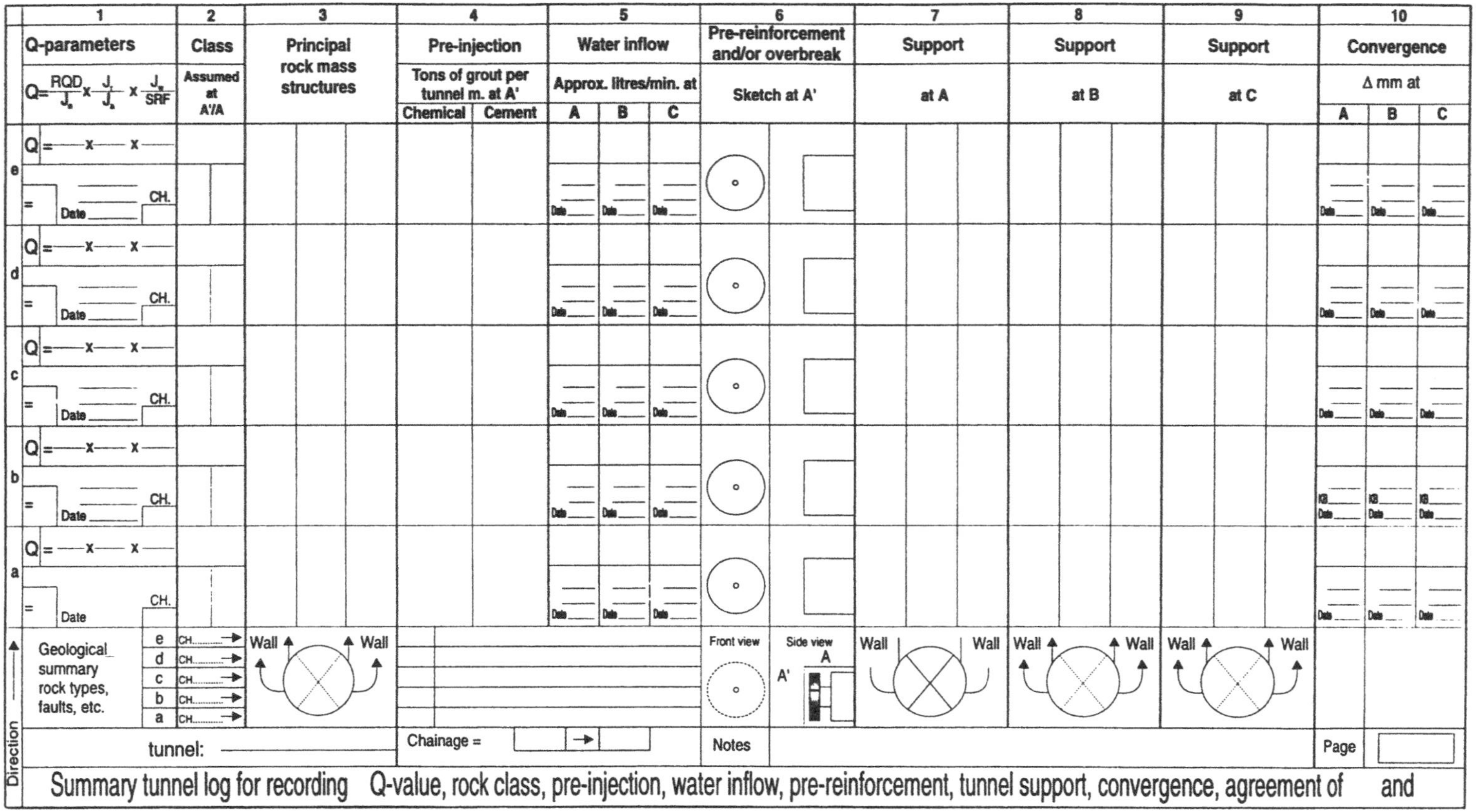
1 Q-parameters $Q=\frac{RQD}{J_n} \times \frac{J_r}{J_a} \times \frac{J_w}{SRF}$
2 Class Assumed at A'/A
3 Principal rock mass structures
4 Pre-injection Tons of grout per tunnel m. at A' Chemical Cement
5 Water inflow Approx. litres/min. at A B C
6 Pre-reinforcement and/or overbreak Sketch at A'
7 Support at A
8 Support at B
9 Support at C
10 Convergence Δ mm at A B C

e, d, c, b, a: Q = — x — x — ; = Date ___ CH.

Geological summary rock types, faults, etc.
e CH. → ; d CH. → ; c CH. → ; b CH. → ; a CH. →
Wall Wall
Front view Side view A A'
Direction
tunnel: ___ Chainage = → Notes Page
Summary tunnel log for recording Q-value, rock class, pre-injection, water inflow, pre-reinforcement, tunnel support, convergence, agreement of and

Figure 66. TBM tunnel log for open, large diameter machines.

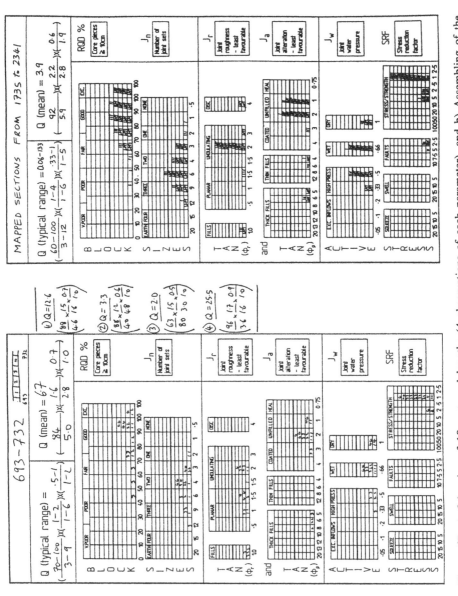

Figure 67. a) *Q*-logging of 10 m tunnel lengths (4 observations of each parameter), and b) Assembling of the *Q*-statistics for 600 m of tunnel (NGI 1998, NGI 1999).

overbreak unless the Q-values are significantly below about 40 in a big tunnel ($\varnothing$ 10 m) or below about 4 in a fairly small tunnel ($\varnothing \approx$ 4 m). There is the added difficulty that an actual Q-value of about 10 in the large tunnel (if the tunnel was drill-and-blasted) may appear to be several times higher in the TBM excavation. In the small tunnel, an actual Q-value of about 1 may appear to be about 4 in the TBM tunnel. This is due to the tendency (actually a correct tendency due to the limited disturbance) for the observer to over-rate RQD and joint spacing and un-der-rate the number of effective joint sets and joint continuity. There will also be potential errors, such as failing to see clay coated joints if overbreak has not oc-curred. These 'central threshold' biases/errors occur in the black-coloured areas shown in Figure 49, as already discussed in Chapter 29.

When comparing Q-logging and RMR logging performed by different parties in the same tunnel, a tendency has been noted for biased RMR observations in the direction of more jointing, because it is easier to record the closer joint spacing and perhaps ignore that much of the tunnel periphery is quite massive. The same could apply to RQD observations, in either the Q-system or RMR method.

A definitive method for avoiding this pitfall is 'histogram logging' as illus-trated in Figure 43 and Figure 67. This method is actually more representative of actual variability, it can take care of the logger's real uncertainty, and it is faster than the 'forced evaluation' of a single value for RQD (or RQD_o), J_n, J_r, J_a, J_w and SRF, which takes significant mental effort in what is often a wet and time-limited environment. The logger may be wet, his paper even wetter and the tunnel invert flowing like a river. The histogram method is the answer here! Later, in the warmth of a site office, with dried logging paper, the engineering geologist can assemble the data, calculate Q-values for individual lengths of the tunnel, and immediately evaluate needs for final support (e.g. at work platforms B and C in Figs 56 and 66).

The logging turnaround time can, and must, be very short, if selection of cor-rect support class and its implementation are to keep pace with the 10 to 100 m per day advance rate. For the engineering geologist the 1 m per day advance with 'unexpected events' (Figs 34 and 35) is the time to log face conditions and super-vise 'pre-reinforcement' measures. In maintenance shifts he will need to follow probe drilling, perhaps interpret sonic logging. There are few idle moments, and each day may have 20 or 24 hrs, meaning two engineering geologists per Owner and two per Contractor. This investment, which also allows interpreted probe drilling, may save long down-periods, which are infinitely more expensive for both parties.

CHAPTER 36

TBM or drill-and-blast excavation?

This chapter heading is posed as a question, and its answer must necessarily precede all the foregoing chapters.

This book is not the place to find TBM rental or purchase costs, but geotechnical aspects such as rate of advance or machine utilisation for different conditions, time periods or lengths of tunnel have been treated in detail.

Knowledge of Q_{TBM}, Q and Q_o actually solves the TBM side of the geotechnical comparison with drill-and-blast, because Q_{TBM} contains machine capabilities (cutter force F) and its comparison with the resistance of the rock mass to cutting (SIGMA$_{cm}$ or SIGMA$_{tm}$ and angle $\beta°$).

Drill-and-blast tunnelling rates and costs are also strongly related to Q-values, based on the assumption that excavation methods and support methods are linked to the Q-system, as they often are in Norway and in increasingly larger numbers of countries around the world.

For a given size of drill-and-blasted tunnel, with a specific end use, the Q-value and ESR value (Fig. 49) will provide suitable permanent support. The support and reinforcement methods listed in this figure detail recommended quantities, based on Q-value ranges.

> *Example*: Span = 10 m, ESR = 1.0 (main road tunnel)
> Q = 1 to 4 (poor)
> Recommended support = B c/c 1.7 to 2.1 m
> S(fr) 5 to 9 cm

The necessary $\approx$ 2 m^3 of shotcrete and 5 rock bolts cost about USD 1000 per meter of tunnel. Adding all the other costs of drilling, explosives, transport, ventilation, mobilisation and profit, the cost of this fully supported 10 m span tunnel may range from about USD 4000/m in the best rock qualities (i.e. $Q \approx$ 10 to 100) to about USD 14,000 in the poorest qualities (i.e. $Q \approx$ 0.01). However, the typical Q-value statistic incorporates a range of Q-values for specific lengths of tunnel. An example from an Oslo TBM tunnel in limestones and shales is given in Figure 68.

In practice, the total cost of the 10 m span tunnel may range from USD 5000 to USD 10,000 due to longer stretches of fair to good rock. In Figure 68, some 4.5

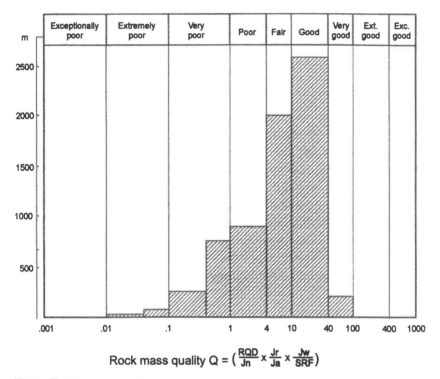

Figure 68. The estimated *Q*-values statistic for a planned tunnel largely determine cost and speed both for TBM and drill-and-blasting (NGI 1999).

km has *Q*-values between 4 and 40, which may bring the total cost closer to the lower limit.

In Figure 34 and 35 (see also Chapters 15, 16 and 17) the decelerating advance rate with time and TBM tunnel length was documented, and also described with simple formulae. Similar sets of data for drill-and-blast tunnels are not readily available (but presumably may exist?). However, drilling a round of 2, 3, 4 or 5 m for the next tunnel advance takes a finite time when there may be 100 or more holes per round and a drill-jumbo with only three booms. A diagram such as Figure 7 for the particular rock mass quality, would give the data necessary to estimate the 'penetration rate' for this partial-cycle of a drill-and-blast tunnel. Each hole can clearly be drilled about 60 times faster than a TBM can penetrate. The total cycle for drill-and-blast tunnels driven in poor rock may consist of:

1. Drilling
2. Charging
3. Blasting
4. Ventilation delay

5. Scaling
6. Shotcreting (temporary) using robot
7. Mucking out
8. Bolting (temporary)

Permanent support will occur outside this cycle.

Since items 1, 2, 5, 6 and 8 will depend strongly on the *Q*-value, it is normal to have a relatively close relationship between the whole cycle time (items 1 to 8 inclusive) and the Q-value. Typical cycle times (hours per round) for a drill-and-blasted tunnel of cross-sectional area 50 m^2 in relation to Q-values would be as follows (Table 20) (based on Grimstad 1999).

As indicated in Figure 69, each rock class in Table 20 has an associated support need, which increases in cost and time of installation as the Q-values reduce below about 30 (the no-support boundary for 8 m span tunnels, Fig. 49).

The 'law of decelerating advance rate' (Chapters 15 and 16) that we see in TBM tunnels is related to geological factors and machine related factors, such as back-up and conveyor performance. In drill-and-blasted tunnels the geological statistic, and its variation with tunnel length is of slightly lower quality than that of the TBM tunnel. However, the impact on schedule of for example $Q < 0.01$, is usually much worse in the TBM tunnel and greater problems and delays may occur, especially if the Q-value is as low as 0.001 (exceptionally poor).

The hypothetical example in Table 21 compares potential advance rates for a

Table 20. Approximate cycle times and advance rates for Norwegian 50 m^2 drill-and-blasted tunnels.

Q-values	0.001-0.01	0.01-0.1	0.1-1	1-10	10-100
Advance/round (m)	1-2	2.5	4	5	5
Cycle time (hrs)	30-23	23-13	13-9	9-5	5-4
Advance rate (m/hr)	0.03-0.09	0.1-0.2	0.3-0.5	0.6-1.0	1.0-1.2
Advance/100 hrs (week)	3-9	10-20	30-50	60-100	100-120

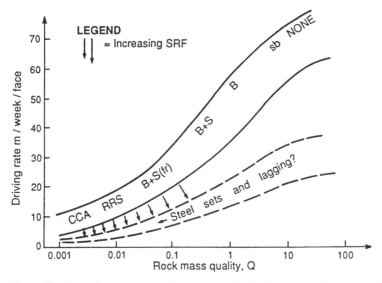

Figure 69. Approximate average advance rates in relation to Q-values, for drill-and-blasted tunnels ranging from 70 to 90 m^2 (Grimstad & Barton 1993).

Table 21. Hypothetical comparison of TBM and drill-and-blast advance rates, assuming $Q_{TBM} \approx Q \approx Q_o$, and 1 week = 100 hours.

$Q \approx Q_o \approx Q_{TBM}$	0.001	0.01	0.1	1.0	10	100	1000
PR m/hr for TBM	(1)	(2)	(4)	5	3	2	1
AR m/hr for TBM	0.015	0.08	0.40	2.08	1.37	0.83	0.38
Assumed m (from Table 11)	−0.9	−0.7	−0.5	−0.22	−0.17	−0.19	−0.21
m/week for TBM	1.5	8	40	208	137	83	38
AR m/hr for D&B	0.03	0.1	0.25	0.55	1.0	1.1	1.1
m/week for D&B	3	10	25	55	100	110	110

Notes: 1) $AR \approx PR \times T^m$ (with $T = 100$ hrs = nominal week). 2) Parentheses indicate operator reduced PR and cutter force.

TBM tunnel and a drill-and-blasted tunnel. In this simplified example, we will assume that $Q_o = Q_{TBM}$.

The drill-and-blasted tunnel may be very competitive if there are many meters of faulted rock and poor ground, and if there is very much extremely good quality, massive rock. According to the example in Table 21, drill-and-blast tunnelling will then be faster. However, if rock quality lies in the middle range and is dominated by Q ($= Q_o$) = 0.1, or 1.0, or 10, the TBM would appear to be two to three times faster than drill-and-blast.

However, there is one common error that has been committed in the example given in Table 21 and that is the limited time frame used for comparison, in this case 1 week, or a nominal 100 hrs. As can be noted from the three 'AR' curves given in Figure 44, and as is reflected in all the case record data given in Figure 35, it is not correct to use a time frame as small as 1 week in the case of TBM tunnels, because the advance rates (and utilisation) decline steadily out to at least 1 year (albeit if the tunnel takes this long to complete). Except in exceptional cases (i.e. best weeks and best months) the *average* AR values will seldom exceed 1.5 m/hr when the time frame is as long as one month. In a long TBM tunnel taking 1 year, the *average* advance rate will usually be less than 1 m/hr. We must therefore conclude that several of the AR values given in Table 21 are overly optimistic, both for TBM (especially) and drill-and-blast (less so).

A nocturne and thorough evaluation is required before choosing either drill-and-blast or TBM for a given project, The common argument that if the tunnel is several kilometers long, TBM will be faster may often be true and sometimes greatly in the TBM's favour. However, the 'law of decelerating advance rate' has so far been difficult to control with yesterday's and today's TBM, due both to the unpredictable nature of geology and hydrogeology and their fault zones, and due to material friction, fatigue and fracture where the TBM and many of the mechanical components are concerned. The predicted Q-value and Q_{TBM} statistic, which should be as reliable as possible, may be the key to the important decision between drill-and-blast and TBM tunnelling.

The final figure (Fig. 70) shows a hypothetical comparison of drill-and-blast and TBM tunnelling based on the example given in Table 21. It emphasises three important points.

1. A false impression of TBM performance is given when only short time-frames are quoted.
2. The steep peak of short-term TBM performance may be particularly sensitive to rock mass conditions.
3. A comparison of rates (TBM and drill-and-blast) is critically dependent on rock conditions, i.e. on the Q-value statistic for the tunnel as a whole, and on Q_{TBM}.

There are obviously important exceptions to the trends shown in Figure 70, because we have set $Q = Q_{TBM}$. A soft, non-abrasive massive rock will have a low value of SIGMA and a lower value of Q_{TBM} than the Q-value suggests, because of more than sufficient thrust. In any case, road-header excavation would be a more suitable method to compare with the TBM under such conditions.

The correct approach will be a separate analysis of the Q-value statistic with drill-and-blast, and a separate analysis of the Q_{TBM} statistic with TBM excavation. Site investigation clearly presents itself as an extremely important investment in deciding between TBM and drill-and-blast, and in correctly planning either method.

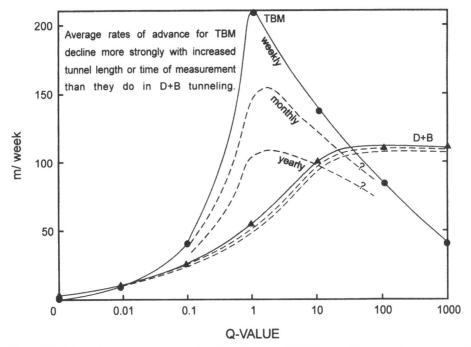

Figure 70. A hypothetical comparison of drill-and-blast and TBM tunnelling rates in m/week (solid curves). Stippled curves represent possible monthly and yearly rates.

CHAPTER 37

Conclusions

1. TBM tunnelling can give truly remarkable tunnel advance rates. An extreme 176 m in a 24-hour day has been recorded, while at the other end of the spectrum, machines can be trapped for years while contracts are renegotiated and alternative tunnelling techniques are sought.

2. Machine-rock interaction in massive, hard, abrasive rock and machine-rock mass interaction in faulted, water-bearing or squeezing conditions give orders of magnitude ranges in advance rate compared to ideal jointed, stable, non-abrasive conditions. Rock mass classification that also gives orders of magnitudes of different qualities is obviously required and has now perhaps been achieved via Q_{TBM}.

3. A slightly modified six-parameter Q-system forms the rock mass descriptive core of Q_{TBM}. Oriented RQD (RQD$_o$ in the tunnelling direction) and J_r/J_a parameters related to the joint set assisting (or hindering) cutter efficiency are all that is changed to obtain Q_o. Parameter ratings are the well known standard values (see Appendix – Section A1.)

4. Machine-rock interaction parameters modify the Q_o-value to account for cutter abrasion (CLI), quartz content, cutter force in relation to an orientation-dependent rock mass strength and the effect of the biaxial stress state on the tunnel face. These 'modifications' are actually multiplications and divisions, and give a potential Q_{TBM} range from about 10^{-3} to 10^9, six orders of magnitude wider than the Q-value range.

5. Due to all the machinery and back-up equipment and delays caused by changed conditions, TBM tunnels show a decelerating advance rate compared to conventional drill-and-blast tunnels as length (or time) increases, following the usual learning curve *increases* in early weeks. Average advance rates per day, per week, per month and per three months show a quite constant deceleration. Extreme value statistics concerning rock mass and hydrogeological conditions must be expected to increase in number as tunnel length (and tunnel depth) increase. This is part of the decelerating advance rate. Only learning curves and greater efficiencies can partly conteract the gradients ($-m$) of decelerating advance.

6. At the lower end of the rock mass quality (Q-value) scale, delays increase in relation to the Q-value, and deceleration gradients are often about –0.5 (at $Q \approx 0.1$), –0.7 (at $Q \approx 0.01$) and even as high as –0.9 (at $Q \approx 0.001$). These very poor rock conditions require probing and pre-treatment, so that the Q-value can be effectively increased and the steep negative gradients reduced. Some hours spent in probe drilling, and some days spent in pre-treatment may save weeks or months of delay in unstable, water-bearing faulted rock.

7. Both the penetration rate PR achieved with uninterrupted boring and the advance rate AR are expressed in terms of m/hr, to facilitate the simple power term formulations with Q_{TBM}. Obviously $AR = U \times PR$ where U = utilisation. However in the present formulations U, which is time dependent, is expressed as T^m, where T is hours of actual time. The deceleration gradient $(-)m$ is based on the real time, since this determines the actual speed of tunnelling in a 24-hour day, 168-hour week or 720-hour month.

8. The simple relationship $PR \approx 5 Q_{TBM}^{-1/5}$ for prediction of penetration rate (which is multiplied by T^m to obtain advance rate AR) can be inverted to the form $Q_{TBM} \approx (PR/5)^5$ for the case of back-analysis of progress. If the intact rock and machine-rock mass interaction parameters (e.g. F/SIGMA) are approximately known, then the above inversion can be used for an approximate back-calculation of the 'Q-value'. Since this will be a strictly oriented Q-value, using RQD_o in the tunnelling direction, it will be more correct to term it Q_o.

9. The gradient $(-)m$ of deceleration is machine dependent, rock matrix dependent and rock mass dependent. The conventional tunnel stability Q-value determines the value of $(-)m$ when the rock mass is of very poor or extremely poor quality ($Q \leq 0.1$) while machine-rock interaction parameters such as cutter life index CLI and quartz content and porosity determine the value of $(-)m$ when the Q-value is at the high quality end of the tunnel stability scale.

10. An assumed quadratic relation between penetration rate and cutter force is achieved by a suitable power term, and the cutter force is normalised by 20 tnf, above which very strong increases in penetration rate are often noted.

11. A key aspect of the Q_{TBM} concept is the comparison of cutter force with an empirical measure of the rock mass resistance to cutter penetration called SIGMA. This is based either on $Q_c^{1/3}$ where $Q_c = \sigma_c / 100 \times Q_o$, or based on $Q_t^{1/3}$ where $Q_t = I_{50}/4 \times Q_o$. Strong I_{50} anisotropy as in slates, and large ratios of σ_c / I_{50} are allowed to influence the value of SIGMA (SIGMA$_{cm}$ or SIGMA$_{tm}$ respectively) which is thereby strongly orientation dependent.

12. TBM tunnelling generally minimises the disturbed zone and reduces the need for tunnel support. This holds for the upper nine orders of magnitude (approx. 1 to 10^9) of Q_{TBM}. At the lowest end of the Q-value and Q_{TBM} scale (approx. 0.1 to 0.001) the effective disturbance may be maximised by the inflexibility of the

TBM and by the adverse effect of the shield on stand-up time limitations. This is especially true when erodible fault gouge and high pressure water are adjacent and unwanted neighbours. Under such conditions it is actually a disadvantage to be using a TBM, and it is inadviseable to ream (or drill-and-blast) a larger diameter tunnel along the same TBM axis. A pilot TBM tunnel in the case of consistently poor rock should perhaps be outside the profile and used for pretreatment of the main tunnels (as successfully achieved at the Channel Tunnel but unsuccessfully attempted at Pinglin where the water flows were too high).

13. In relation to the Q-system tunnel support chart and the line separating supported and unsupported excavations, a TBM tunnel will give an apparent (and partially real) increase in the Q-value of about 2 to 5 times in this region. This is where the TBM tunnel support needs are much reduced. When the Q-value is lower than in the *central threshold* area, the TBM tunnel will show similar levels of overbreak or instability as the drill-and-blasted tunnel, and Q-system derived final support solutions can apply. However, they may be preceded by non-reinforcing temporary steel sets and lagging (and void formation) each of which require due consideration when finalising support.

14. Due to the especially adverse consequences of bad ground (fault zones, flooding, squeezing, etc.) in TBM tunnelling, it is especially advisable to use 'systematic-when-needed' probe drilling, core drilling or remote sensing, and have the facilities for systematic and efficient pre-injection, spiling and shotcrete application, if necessary performed from ahead of the cutterhead. The delays involved in each of these measures for reducing risk will usually pay off handsomely, because they give the opportunity ahead of time to improve conditions (raise Q and Q_{TBM} magnitudes) and thereby create less damage to advance rate schedules, by only penetrating pre-treated extreme ground.

15. The expectation of less dependence on rock mass conditions when using shielded and telescopic double shielded machines may not be realised as often as expected. There is convincing evidence to suggest similar dependence of advance rates on rock mass quality for both open and shielded machines, which suggests that lack of access to the rock for early support can outweigh the assumed advantages of the shielded machines when stand-up time is limited. Shields and double shields are often 'too long' in relation to the reduced stand-up time and advance rate reductions, and rock conditions can then deteriorate under poorly controlled conditions.

16. Longitudinal profiles along planned tunnel axes should provide the necessary statistics for Q and Q_{TBM}, using a combination of geological mapping, drill core and seismic velocities from the site investigations. Each will be corrected and updated during tunnelling, or improved using horizontal drillcore. The Q and Q_{TBM} statistics will make comparison of TBM and drill-and-blast tunnelling rates a quantitative exercise, justifying investment in improved site investigations.

Appendix

This appendix contains the final version of the Q_{TBM} diagram, and a table containing all the six Q parameter ratings. Input data needs are summarised, together with a listing of key equations. Finally, two work sheets are presented (blank and filled in) to show how the data can be assembled and how calculations can be readily made by hand, when few zones are to be calculated (feasibility stage). A computer program has been written to ease calculation when many domains are involved, and for project follow-up (www.Qtbm.com).

A1 Q-METHOD OF ROCK MASS CLASSIFICATION

These tables contain all the ratings necessary for estimating the Q-value of a rock mass. The ratings form the basis for both the Q and Q_{TBM} estimates.

$$Q = \frac{RQD}{J_n} \times \frac{J_r}{J_a} \times \frac{J_w}{SRF}$$

1. ROCK QUALITY DESIGNATION		RQD (%)
A	Very poor.	0-25
B	Poor.	25-50
C	Fair.	50-75
D	Good.	75-90
E	Excellent.	90-100

Notes: – Where RQD is reported or measured as ≤10 (including 0), a nominal value of 10 is used to evaluate Q.
 – RQD intervals of 5, i.e. 100, 95, 90, etc. are sufficiently accurate.

2. JOINT SET NUMBER		J_n
A	Massive, no or few joints.	0.5-1
B	One joint set.	2
C	One joint set plus random joints.	3
D	Two joint sets.	4
E	Two joint sets plus random joints.	6
F	Three joint sets.	9
G	Three joint sets plus random joints.	12
H	Four or more joint sets, random, heavily jointed, 'sugar-cube', etc.	15
J	Crushed rock, earthlike.	20

Notes: – For tunnel intersections, use $(3.0 \times J_n)$.
 – For portals use $(2.0 \times J_n)$.

3. JOINT ROUGHNESS NUMBER		J_r
a) *Rock-wall contact*, and b) *Rock-wall contact before 10 cm shear*		
A	Discontinuous joints.	4
B	Rough or irregular, undulating.	3
C	Smooth, undulating.	2
D	Slickensided, undulating.	1.5
E	Rough or irregular, planar.	1.5
F	Smooth, planar.	1.0
G	Slickensided, planar.	0.5

Notes: – Descriptions refer to small-scale features and intermediate scale features, in that order.

c) *No rock-wall contact when sheared*		J_r
H	Zone containing clay minerals thick enough to prevent rock-wall contact.	1.0
J	Sandy, gravely or crushed zone thick enough to prevent rock-wall contact.	1.0

Notes: – Add 1.0 if the mean spacing of the relevant joint set is greater than 3 m.

– $J_r = 0.5$ can be used for planar, slickensided joints having lineations, provided the lineations are oriented for minimum strength.

– J_r and J_a classification is applied to the joint set or discontinuity that is least favourable for stability both from the point of view of orientation and shear resistance, τ (where $\tau \approx \sigma_n$ tan (J_r / J_a).

4. JOINT ALTERATION NUMBER		ϕ_r approx.	J_a
a) *Rock-wall contact (no mineral fillings, only coatings)*			
A	Tightly healed, hard, non-softening, impermeable filling, i.e. quartz or epidote.	–	0.75
B	Unaltered joint walls, surface staining only.	25-35°	1.0
C	Slightly altered joint walls. Non-softening mineral coatings, sandy particles, clay-free disintegrated rock, etc.	25-30°	2.0
D	Silty- or sandy-clay coatings, small clay fraction (non-softening).	20-25°	3.0
E	Softening or low friction clay mineral coatings, i.e. kaolinite or mica. Also chlorite, talc, gypsum, graphite, etc. and small quantities of swelling clays.	8-16°	4.0
b) *Rock-wall contact before 10 cm shear (thin mineral fillings)*			
F	Sandy particles, clay-free disintegrated rock, etc.	25-30°	4.0
G	Strongly over-consolidated non-softening clay mineral fillings (continuous, but <5 mm thickness).	16-24°	6.0
H	Medium or low over-consolidation, softening, clay mineral fillings (continuous, but <5 mm thickness).	12-16°	8.0
J	Swelling-clay fillings, i.e. montmorillonite (continuous, but <5 mm thickness). Value of J_a depends on percentage of swelling clay-size particles, and access to water, etc.	6-12°	8-12
c) *No rock-wall contact when sheared (thick mineral fillings)*			
K,L,M	Zones or bands of disintegrated or crushed rock and clay (see G, H and J for description of clay condition).	6-24°	6, 8 or 8-12
N	Zones or bands of silty- or sandy-clay, small clay fraction (non-softening).	–	5.0
O,P,R	Thick, continuous zones or bands of clay (see G, H and J for description of clay condition).	6-24°	10, 13 or 13-20

5. JOINT WATER REDUCTION FACTOR		Approx. water pressure (kg/cm^2)	J_w
A	Dry excavations or minor inflow, i.e. < 5 l/min. locally.	<1	1.0
B	Medium inflow or pressure, occasional out-wash of joint fillings.	1-2.5	0.66
C	Large inflow or high pressure in competent rock with unfilled joints.	2.5-10	0.5
D	Large inflow or high pressure, considerable outwash of joint fillings.	2.5-10	0.33
E	Exceptionally high inflow or water pressure at blasting, decaying with time.	>10	0.2-0.1
F	Exceptionally high inflow or water pressure continuing without noticeable decay.	>10	0.1-0.05

Notes: – Factors C to F are crude estimates. Increase J_w if drainage measures are installed.
  – Special problems caused by ice formation are not considered.

6. STRESS REDUCTION FACTOR

a) *Weakness zones intersecting excavation, which may cause loosening of rock mass when tunnel is excavated*		SRF
A	Multiple occurrences of weakness zones containing *clay* or chemically disinte-grated rock, very loose surrounding rock (any depth).	10
B	Single weakness zones containing *clay* or chemically disintegrated rock (depth of excavation ≤50 m).	5
C	Single weakness zones containing *clay* or chemically disintegrated rock (depth of excavation >50 m).	2.5
D	Multiple shear zones in competent rock (*clay-free*), loose surrounding rock (any depth).	7.5
E	Single shear zones in competent rock (*clay-free*), (depth of excavation ≤50 m).	5.0
F	Single shear zones in competent rock (*clay-free*), (depth of excavation >50 m).	2.5
G	Loose, open joints, heavily jointed or 'sugar cube', etc. (any depth).	5.0

Notes: – Reduce these values of SRF by 25-50% if the relevant shear zones only influence but do not intersect the excavation.

b) *Competent rock, rock stress problems*		σ_c/σ_1	σ_θ/σ_c	SRF
H	Low stress, near surface, open joints.	>200	<0.01	2.5
J	Medium stress, favourable stress condition.	200-10	0.01-0.3	1
K	High stress, very tight structure. Usually favourable to stability, may be unfavourable for wall stability.	10-5	0.3-0.4	0.5-2
L	Moderate slabbing after >1 hour in *massive* rock.	5-3	0.5-0.65	5-50
M	Slabbing and rock burst after a few minutes in *massive* rock.	3-2	0.65-1	50-200
N	Heavy rock burst (strain-burst) and immediate dy-namic deformations in *massive* rock.	<2	>1	200-400

Notes: – For strongly anisotropic virgin stress field (if measured): When $5 \leq \sigma_1/\sigma_3 \leq 10$, reduce σ_c to 0.75 σ_c. When $\sigma_1/\sigma_3 > 10$, reduce σ_c to 0.5 σ_c, where σ_c = unconfined compression strength, σ_1 and σ_3 are the major and minor principal stresses, and σ_θ = maximum tangential stress (estimated from elastic theory).
  – Few case records available where depth of crown below surface is less than span width. Suggest SRF increase from 2.5 to 5 for such cases (see H).

c) *Squeezing rock: plastic flow of incompetent rock under the influence of high rock pressure*		σ_θ/σ_c	SRF
O	Mild squeezing rock pressure.	1-5	5-10
P	Heavy squeezing rock pressure.	> 5	10-20

Notes: – Cases of squeezing rock may occur for depth $H > 350\,Q^{1/3}$ (Singh et al., 1992). Rock mass compression strength may be estimated from $q \approx 0.7\,\gamma\,Q^{1/3}$ (MPa) where γ = rock density in kN/m^3 (Singh, 1993).

d) *Swelling rock: chemical swelling activity depending on presence of water*		SRF
R	Mild swelling rock pressure	5-10
S	Heavy swelling rock pressure	10-15

A2 Q_{TBM} – THE FINAL VERSION OF FIGURE 44

The parameters used to define Q_{TBM} have been developed during a process of trial and error using case records. This process was also used when developing the Q-value twenty-five years earlier. The process is not complete and future refinements of this diagram are expected (see Fig. A1).

A3   INPUT DATA SUMMARY FOR ESTIMATING PR AND AR USING Q_{TBM}

This section contains a summary of the equations and input data for estimating PR and AR using Q_{TBM}. The following equations are needed to evaluate gradient (m) and Q_{TBM}. It will be noted that one or two subscripts have been added to original equations (e.g. Q_o, s, c) to give consistency in evaluation and small improvements (see Equations 7 and 19).

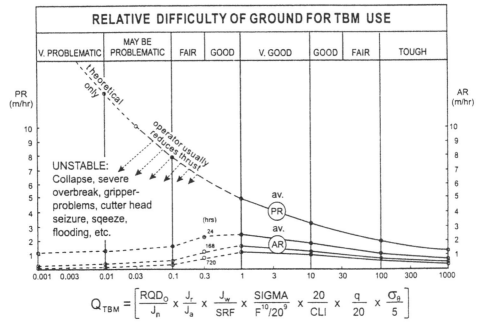

$$Q_{TBM} = \left[\frac{RQD_0}{J_n} \times \frac{J_r}{J_a} \times \frac{J_w}{SRF} \times \frac{SIGMA}{F^{10}/20^9} \times \frac{20}{CLI} \times \frac{q}{20} \times \frac{\sigma_\theta}{5} \right]$$

Figure A1. Q_{TBM} – The final version of Figure 44.

A) $$Q = \frac{RQD}{J_n} \times \left(\frac{J_r}{J_a}\right)_s \times \frac{J_w}{SRF}$$

Conventional Q-parameters for estimating *stability* and gradient m_1 (e.g. Fig. 45 and Table 11) and tunnel support needs (e.g. Figs 49, 55, 56 and Table 18).

B) $$Q_o = \frac{RQD_o}{J_n} \times \left(\frac{J_r}{J_a}\right)_c \times \frac{J_w}{SRF}$$

RQD_o is oriented in tunnelling direction, J_r and J_a are chosen for the joint set or discontinuity most assisting (or hindering) *cutter* penetration.

C) $$Q_{TBM} = Q_o \times \frac{SIGMA}{F^{10}/20^9} \times \frac{20}{CLI} \times \frac{q}{20} \times \frac{\sigma_\theta}{5} \qquad \text{(Eq. 28)}$$

1) $SIGMA = SIGMA_{cm} = 5\gamma Q_c^{1/3}$ where $Q_c = Q_o \times \dfrac{\sigma_c}{100}$ (modified Eq. 7)

2) $SIGMA = SIGMA_{tm} = 5\gamma Q_c^{1/3}$ where $Q_t = Q_o \times \dfrac{I_{50}}{4}$ (modified Eq. 19)

D) $$m = m_1 \left(\frac{D}{5}\right)^{0.20} \left(\frac{20}{CLI}\right)^{0.15} \left(\frac{q}{20}\right)^{0.10} \left(\frac{n}{2}\right)^{0.05} \qquad \text{(Eq. 24)}$$

A total of twenty basic parameters or dimensions are required, many of which can be estimated by experienced engineering geologists during preliminary site evaluation. We have:
- The nine essential rock mass parameters: RQD, RQD_o J_n, (J_r/J_a) stability, (J_r/J_a) cutter penetration, J_w, and SRF (see Q-tables in Section A1).
- The four essential rock and rock mass strength parameters for estimating the anisotropic strength: σ_c, (where relevant $\sigma_c\bot$, $\sigma_c\|$), I_{50} (where relevant $I_{50}\bot$, $I_{50}\|$), $\beta°$, and γ.
- The three essential rock abrasion parameters: CLI, q, and n.
- The four fundamental dimensions: diameter = D, length = L, time = T, and cutter force = F.

The solutions obtained for Q_{TBM} and m are substituted in Equations 11 and 12 to obtain the required estimates of PR and AR.

$$PR \approx 5 Q_{TBM}^{-1/5} \qquad \text{(Eq. 11)}$$

$$AR \approx 5 Q_{TBM}^{-1/5} T^m \qquad \text{(Eq. 12)}$$

The time (T) for completion of a given length (L) of tunnel, or length of specific geological zone or length of geotechnical domain is estimated using Equation 16:

$$T = \left(\frac{L}{PR}\right)^{\frac{1}{1+m}} \qquad \text{(Eq. 16)}$$

using subscripts such as L_{z1}, L_{z2} for the zones, and L_{D1}, L_{D2} for the domains.

In Figure A2 (after Nelson et al. 1999), the above equations would need to be evaluated for each domain. In the lower diagram the cell with address 1.5 (zone 1, domain 5) is highlighted. A domain with address and length $L_{2,3}$ would be from zone 2 and refer to domain 3. The first task of an engineering geologist would be to locate these domains on a vertical section along the tunnel, based on all available information.

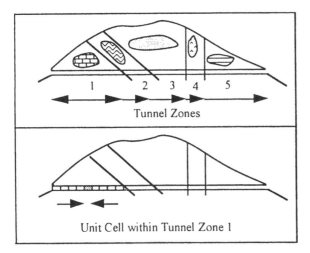

Figure A2. Zones (of different geology) and cells (of different geotechnical domain) should form the basis for Q_{TBM} estimation (extract from Nelson et al., 1999.)

When using the Q-system to describe rock stability conditions and Q_o to describe TBM penetration potential, it is clear that *each tunnel, each zone* and *each domain* will have its individual Q-parameter statistics. An example of a whole-tunnel Q-statistic was given in Figure 68. This could be a starting point for estimation of variations in Q_{TBM}.

At the more detailed *zone* scale, the Q-parameter statistics shown in Figure 67b could be an illustrative example, based on surface mapping, borehole data, etc. At the most detailed *domain* scale, the individual logging (or estimation) of geotechnical domains will be the correct unit. In Figure 67a, the tunnel has been logged at 10 m intervals. This uniformity of chainage (10 m) will seldom be suitable for *domain* logging or estimation, which should be driven instead by the local rock mass, by geological and hydrogeological conditions, which we have termed the geotechnical *domains*.

A4   WORKED EXAMPLE

In the following work sheets, the *input data* (Sheet I) and the *performance estimation* (Sheet II) have been filled in for the hypothetical case of a 16 km long tunnel with equal lengths (4 km) of sandstone (massive, abrasive), phyllite (ideally jointed), mica schist (ideally jointed) and granite (massive, very strong).

The Q_{TBM} method is shown to estimate extremely favourable rates of advance for the phyllite and schist, but poor rates of advance for the sandstone and granite.

INPUT DATA (SHEET I)

A) STABILITY (AND GRADIENT m_1) $\quad Q = \left(\dfrac{RQD}{J_n}\right) \times \left(\dfrac{J_r}{J_a}\right)_s \times \left(\dfrac{J_w}{SRF}\right)$ (least favourable for *stability*)

Zone	m_1	RQD/J_n	J_r/J_a	J_w/SRF	Q
1 Sandstones	−0.17	100/9	2/1	0.5/1	11
2 Phyllites	−0.19	35/9	1.5/1	1.0/1	6
3 Mica schists	−0.20	50/9	1.0/1	0.66/1	4
4 Granites	−0.18	100/6	2/1	0.66/1	22

B) ORIENTED Q_o (IN TUNNELLING DIRECTION) $\quad Q_o = \left(\dfrac{RQD_o}{J_n}\right) \times \left(\dfrac{J_r}{J_a}\right)_c \times \left(\dfrac{J_w}{SRF}\right)$ (most affecting *cutters*)

Zone	$\beta°$	RQD_o/J_n	J_r/J_a	J_w/SRF	Q_o
1 Sandstones	20/70	100/9	2/1	0.5/1	11
2 Phyllites	60	30/9	1.5/1	1.0/1	5
3 Mica schists	60	45/9	1.0/1	0.66/1	3
4 Granites	10/80	100/6	2/1	0.66/1	22

C) ROCK MASS STRENGTH (SIGMA) $\quad SIGMA_{cm} = 5\gamma Q_c^{1/3}$, where $\left(Q_c = Q_o \times \dfrac{\sigma_c}{100}\right)$

$$SIGMA_{tm} = 5\gamma Q_t^{1/3}, \text{where} \left(Q_t = Q_o \times \dfrac{I_{50}}{4}\right)$$

Zone	γ	σ_c	Q_c	$SIGMA_{cm}$ (MPa)	I_{50}	Q_t	$SIGMA_{tm}$ (MPa)
1 Sandstones	2.5	125	14	30	5	–	–
2 Phyllites	2.6	75	4	21	1	1.25	14
3 Mica schists	2.6	150	4.5	21	4	3.0	19
4 Granites	2.7	200	44	48	8	–	–
		(∥ or ⊥)			(∥ or ⊥)		

D) Q_{TBM} $\quad Q_{TBM} = Q_o \times \dfrac{SIGMA}{F^{10}/20^9} \times \dfrac{20}{CLI} \times \dfrac{q}{20} \times \dfrac{\sigma_\theta}{5}$ $\left(\times 20^9/F^{10} = 0.0054 \text{ with } 25 \text{ tnf/cutter}\right)$

Zone	Q_o	SIGMA (MPa)	F (tnf)	CLI	q (%)	σ_θ (MPa)	Q_{TBM}
1 Sandstones	11	30	25	10	70	8	20
2 Phyllites	5	14	25	20	20	8	0.6
3 Mica schists	3	19	25	15	20	8	0.7
4 Granites	22	48	25	10	35	12	48

E) GRADIENT m $\quad m = m_1 \left(\dfrac{D}{5}\right)^{0.20} \times \left(\dfrac{20}{CLI}\right)^{0.15} \times \left(\dfrac{q}{20}\right)^{0.10} \times \left(\dfrac{n}{2}\right)^{0.05}$

Zone	m_1	D (m)	n (%)	m
1 Sandstones	−0.17	10.7	15	−0.27
2 Phyllites	−0.19	10.7	5	−0.23
3 Mica schists	−0.20	10.7	2	−0.24
4 Granites	−0.18	10.7	1	−0.24

CALCULATION (SHEET II)

F) PENETRATION RATE $\quad$ PR $\approx 5Q_{TBM}^{-1/5}$ $\qquad$ AR $= $ PR $\times T^{m}$

Zone	Q_{TBM}	PR (m/hr)	AR (m/hr)
1 Sandstones	20	2.7	0.18
2 Phyllites	0.6	5.5	0.77
3 Mica schists	0.7	5.4	0.67
4 Granites	48	2.3	0.22

G) TIME TO ADVANCE LENGTH L $\qquad T = \left(\dfrac{L}{PR}\right)^{\frac{1}{1+m}}$

Zone	L (m)	m	$\left(\dfrac{1}{1+m}\right)$	T (hr)	$T \times $ AR $= L$*	Assume max. 8736 hrs/yr
1 Sandstones	4000	−0.27	1.37	22,070	4002	2.53 yrs
2 Phyllites	4000	−0.23	1.30	5,250	4026	0.60 yrs
3 Mica schists	4000	−0.24	1.32	6,140	4087	0.70 yrs
4 Granites	4000	−0.24	1.32	18,933	4097	2.17 yrs

$\Sigma L = 16,000$ (m), $\qquad \Sigma T = 52,393$ (hrs) $\qquad = (6.00$ yrs).
*Rough check of AR and T (errors due to rounding).

H) OVERALL PERFORMANCE $\qquad$ PR (weighted mean), ΣT, ΣL $\qquad \overline{PR} = \left(\dfrac{PR_1 L_1 + PR_2 L_2 \text{ etc.}}{L_1 + L_2 \text{ etc.}}\right)$

$$\overline{AR} = \left(\dfrac{AR_1 L_1 + AR_2 L_2 \text{ etc.}}{L_1 + L_2 + L_3 \text{ etc.}}\right)$$

Zones	ΣL (m)	ΣT (hr)	$\overline{PR}$	$\overline{AR}$
1, 2, 3 and 4	16,000	52,393	$\dfrac{63,600}{16,000} = 3.98$	$\dfrac{7,280}{16,000} = 0.46$

I) AR AT THE END OF THE PROJECT $\quad$ AR(end) $= \overline{PR} \times \Sigma T^{\overline{m}}$

Zones	$\overline{m}$	$\overline{PR}$	$\overline{AR}$	AR (end)
1, 2, 3 and 4	0.245	3.98	0.46	0.28

The ideal tunnelling predicted in the phyllites and schists clearly indicates the great benefit of TBM tunnelling. In this example the massive, hard-to-bore sandstones and granites occur at either end of the tunnel, and could be drilled-and-blasted while waiting for TBM delivery.

INPUT DATA (SHEET I)

A) STABILITY (AND GRADIENT m_1) $\quad Q = \left(\dfrac{RQD}{J_n}\right) \times \left(\dfrac{J_r}{J_a}\right)_s \times \left(\dfrac{J_w}{SRF}\right)$ (least favourable for *stability*)

Zone	m_1	RQD/J_n	J_r/J_a	J_w/SRF	Q
1					
2					
3					
4					

B) ORIENTED Q_o (IN TUNNELLING DIRECTION) $\quad Q_o = \left(\dfrac{RQD_o}{J_n}\right) \times \left(\dfrac{J_r}{J_a}\right)_c \times \left(\dfrac{J_w}{SRF}\right)$ (most affecting *cutters*)

Zone	$\beta°$	RQD_o/J_n	J_r/J_a	J_w/SRF	Q_o
1					
2					
3					
4					

C) ROCK MASS STRENGTH (SIGMA)

$$SIGMA_{cm} = 5\gamma Q_c^{1/3} \quad SIGMA_{tm} = 5\gamma Q_t^{1/3} \quad \left(Q_c = Q_o \times \frac{\sigma_c}{100}\right)\left(Q_t = Q_o \times \frac{I_{50}}{4}\right)$$

Zone	γ	σ_c	Q_c	$SIGMA_{cm}$ (MPa)	I_{50}	Q_t	$SIGMA_{tm}$ (MPa)
1							
2							
3							
4							
		(‖ or ⊥)			(‖ or ⊥)		

D) Q_{TBM} $\quad Q_{TBM} = Q_o \times \dfrac{SIGMA}{F^{10}/20^9} \times \dfrac{20}{CLI} \times \dfrac{q}{20} \times \dfrac{\sigma_\theta}{5}$ $\quad \left(\times 20^9/F^{10} = 0.0054 \text{ with } 25\,tnf/cutter\right)$

Zone	Q_o	SIGMA (MPa)	F (tnf)	CLI	q (%)	σ_θ (MPa)	Q_{TBM}
1							
2							
3							
4							

E) GRADIENT m $\quad m = m_1 \left(\dfrac{D}{5}\right)^{0.20} \times \left(\dfrac{20}{CLI}\right)^{0.15} \times \left(\dfrac{q}{20}\right)^{0.10} \times \left(\dfrac{n}{2}\right)^{0.05}$

Zone	m_1	D	n (%)	m
1				
2				
3				
4				

CALCULATION (SHEET II)

F) Penetration rate $\quad PR \approx 5Q_{TBM}^{-1/5} \qquad AR = PR \times T^m$

Zone	Q_{TBM}	PR (m/hr)	AR (m/hr)
1			
2			
3			
4			

G) Time to advance length $L \qquad T = \left(\dfrac{L}{PR}\right)^{\frac{1}{1+m}}$

Zone	L (m)	m	$\left(\dfrac{1}{1+m}\right)$	T (hr)	$T \times AR = L^*$	Assume max. 8736 hrs/yr
1						
2						
3						
4						

H) Overall performance $\quad$ PR (weighted mean), $\Sigma T, \Sigma L \qquad \overline{PR} = \left(\dfrac{PR_1 L_1 + PR_2 L_2 \text{ etc.}}{L_1 + L_2 \text{ etc.}}\right)$

$$\overline{AR} = \left(\frac{AR_1 L_1 + AR_2 L_2 \text{ etc.}}{L_1 + L_2 + L_3 \text{ etc.}}\right)$$

Zones	ΣL (m)	ΣT (hr)	$\overline{PR}$ (m/hr)	$\overline{AR}$ (m/hr)
1 to 4				
5 to 8				
9 to 12				
13 to 16				

I) AR at end of project $\quad AR(\text{end}) = \overline{PR} \times \Sigma T^{\overline{m}}$

Zones	$\overline{m}$	$\overline{PR}$	$\overline{AR}$	AR (end)
1 to 4				
5 to 9				
8 to 12				
13 to 16				

Note: Averaging PR and AR and m by length L is shown in the above examples. Averaging by time T may be a valid alternative. This option is given in the Q_{TBM} computer program. When fault zones cause large delays, averaging by time T gives a more conservative end result.

References

Aasen, O. 1980. Influence of rock properties and jointing by disc cutting with a triple-boom tunnel boring machine. [In Norwegian.] Fjellsprengningsteknikk. Bergmekanikk/Geoteknikk. 27.1-27.9. Trondheim: Tapir Press.

Aeberli, H.U. & Wanner, H. 1978. On the influence of discontinuities at the application of tunnelling machines. *Proc. 3rd IAEG congress, Madrid.* Part III(2): 7-14. Imprime ADOSA.

Alber, M. 1996. Prediction of penetration and utilization for hard rock TBMs. In Barla (ed.), *Eurock '96.* 721-725. Rotterdam: Balkema.

Andreis, I. & Valent, G. 1993. Yindaruqin Irrigation Project: High speed tunnelling in China. In Bowerman & Monsees (eds), *Proc. RETC. Boston, MA.* 391-399. Littleton, CO: Soc. for Mining, Metallurgy, and Exploration, Inc.

Asting, G. 1981. Experiences with pregrouting in sewage tunnels in the Oslo area. *Norwegian Tunnelling Technology Publication No. 2:* 57-63. Trondheim: Tapir Press.

Astolfi, G., Sapigni, M., Barla, G. & Innaurato, N. 1999. Inclines: some rock mechanics and excavation problems in two Italian cases. *Proc. 9th ISRM congress, Paris.* 1: 253-258. Rotterdam: Balkema.

Bandis, S. 1990. Fracture modes around tunnels in weak rocks. *Terzo Cirlo di Conferenze di Meccanica e Ingegneria delle Rocce*, Politecnio di Torino, 1990. Padova: SGE Editoriali.

Barton, N. 1986. Deformation phenomena in jointed rock. 8th Laurits Bjerrum Memorial Lecture, Oslo. *Géotechnique.* 36(2): 147-167. London: Institution of Civil Engineers.

Barton, N. 1991. Geotechnical design. *World Tunnelling*, November 1991. 410-416.

Barton, N. 1995. The influence of joint properties in modelling jointed rock masses. In Fuji (ed.), *Keynote Lecture – Proc. 8th ISRM congress, Tokyo.* 3: 1023-1032. Rotterdam: Balkema.

Barton, N. 1996. Rock mass characterization and seismic measurements to assist in the design and execution of TBM projects. *Keynote Lecture – Proc. of 1996 Taiwan Rock Engineering Symposium.* 1-16. Taipei: National Taiwan University Centre for Education.

Barton, N. 1998. Quantitative description of rock masses for the design of NMT reinforcement. *Keynote Lecture – Int. Conf. on Hydro Power Development in Himalayas, Shimla, India.* 379-400. New Dehli: Oxford and IBH Publishing Co. PVT. Ltd.

Barton, N. 1999. Rock mass characterization from seismic measurements. Keynote Lecture – SAROCKS '98, 2nd Brazilian Symp., 5th South American Rock Mechanics Congress, Santos, Brazil. [Unpublished].

Barton, N. 2000. Seismic velocity and rock quality. [In press].

Barton, N. & Bakhtar, K. 1983. Instrumentation and analysis of a deep shaft in quartzite. In Mathewson (ed.), *24th US symposium on rock mechanics, Texas A&M Univ.* 371-384. College Station, TX: Association of Engineering Geologists.

Barton, N. & Bandis, S.C. 1990. Review of predictive capabilities of JRC-JCS model in engineering practice. *Proc. Int. Symp. on Rock Joints, Loen, Norway.* 603-610. Rotterdam: Balkema.

Barton, N. & Grimstad, E. 1994. The Q-system following twenty years of application in NMT support selection, 43rd Geomechanic Colloquy, Salzburg. *Felsbau*, 6/94. 428-436.

Barton, N. & Warren, C. 1995. Rock mass classification of chalk marl in the UK Channel Tunnels, Channel Tunnel Engineering Geology Symposium, Brighton, September 1995. [Unpublished].

Barton, N., Lien, R. & Lunde, J. 1974. Engineering classification of rock masses for the design of tunnel support. *Rock Mechanics* 6(4): 189-236. Springer-Verlag.

Barton, N., Lien, R., Løset, F., Løken, T., Grimstad, E., Hansteen, H., Hårvik, L. & Christiansson, M. 1986. Methods for selecting support in sub-sea rock tunnels. *Proc. of Int. Symp. Strait Crossings, Stavanger, Norway.* Trondheim: Tapir Press.

Barton, N., By, T-L., Chryssanthakis, P., Tunbridge, L., Kristiansen, J., Løset, F., Bhasin, R.K., Westerdahl, H. & Vik, G. 1994. Predicted and measured performance of the 62 m span Norwegian Olympic Ice Hockey Cavern at Gjøvik. *Int. J. Rock Mech. Min. Sci. & Geomech. Abstr.* 31(6): 617-641. UK: Pergamon.

Bieniawski, Z.T. 1989. Engineering rock mass classifications: A complete manual for engineers and geologists in mining, civil and petroleum engineering. 251 p. J. Wiley.

Blindheim, O.T. 1987. Tunnel boring or drill and blast – geological factors to consider – case analyses. *Proc. Underground Hydropower Plants, Oslo.* 333-340. Trondheim: Tapir Press.

Blindheim, O.T., Dahl Johansen, E. & Johannessen, O. 1979. Criteria for the selection of fullface tunnel boring or conventional tunnelling. *Proc. 4th ISRM congress, Montreux.* 4: 341-346. Rotterdam: Balkema.

Bradley, W.B. 1978. Failure of inclined boreholes. Transactions of the American Society of Mechanical Engineers. 101: 232-239.

Bruland, A., Dahlø, T.S.& Nilsen, B. 1995. Tunnelling performance estimation based on drillability testing. In Fuji (ed.), *Proc. 8th ISRM congress, Tokyo.* 1: 123-126. Rotterdam: Balkema.

Calin, L. & Fernandez, J. 1997. TARP – Comparative study of the mining and concrete lining operations for tunnels with diameters over 9 m (30 ft). In Carlson & Budd (eds), *Proc. RETC. Las Vegas, NA.* Littleton, CO: Soc. for Mining, Metallurgy, and Exploration, Inc.

Chen, C.-N. & Guo, G.-C. 1997. Rock mass classification and guideline for tunnel convergence. *Journal of the Chinese Institute of Civil and Hydraulic Engineering, Taiwan.* 9(3): 359-367.

Chavan Jr, F., Ritz, W., Windler, H.J. & Yanagisawa, S. 1989. Zurichberg railroad tunnel. In Pond & Kenney (eds), *Proc. RETC, Los Angeles CA.* 663-677. Littleton, CO: Soc. for Mining, Metallurgy, and Exploration, Inc.

Chryssanthakis, P. 1991. NGI contract report.

Chryssanthakis, P., Barton, N. Lorig, L., Christianson, M. 1997. Numerical simulation of fibre reinforced shotcrete in a tunnel using the Discrete Element Method. In Kim (ed.), *Proc. of NY Rocks '97; Linking Science to Engineering.* 845-854. Also: *Int. J. Rock Mech and Min Sci.* 34: 3-4.

Corry, T.B. 1995. Construction of the 1122/1130 tunnels on the Superconducting Super Collider project. In Williamson & Gowring (eds), *Proc. RETC. San Francisco, CA.* 409-423. Littleton, CO: Soc. for Mining, Metallurgy, and Exploration, Inc.

Cundall, P.A. & Hart, R.D. 1993. Numerical modeling of discontinua. In Hudson et al. (eds), *Comprehensive rock engineering.* Ch. 9(2): 231-243.

Dalton, F.E., DiVita, L.R. & Macaitis, W.A. 1993. TARP tunnel boring machine performance Chicago. In Bowerman & Monsees (eds), *Proc. RETC. Boston, MA.* 445-451. Littleton, CO: Soc. for Mining, Metallurgy, and Exploration, Inc.

Davey, G.M., Dickson, K.R. & Gowring, I.M. 1991. Kemano power tunnel. In Wightman & McCarry (eds), *Proc. RETC. Seattle, WA.* 487-505. Littleton, CO: Soc. for Mining, Metallurgy, and Exploration, Inc.

Deere, D.W., Spitzer, R.H., Ozdemir, L. & Blyler, J.B. 1995. Conditions encountered in the construction of Stanley Canyon Tunnel Colorado Springs, Colorado. In Williamson & Gowring (eds), *Proc. RETC. San Francisco, CA.* 3-20. Littleton, CO: Soc. for Mining, Metallurgy, and Exploration, Inc.

Deering, K., Dolligner, G.L., Krauter, D. & Roby, J.A. 1991. Development and performance of large diameter cutters for use on high performance TBM's. In Wightman & McCarry (eds), *Proc. RETC. Seattle, WA.* 807-814. Littleton, CO: Soc. for Mining, Metallurgy, and Exploration, Inc.

Deva, Y., Dayal, H.M. & Mehrotra, A. 1994. Artesian blowout in a TBM driven water conductor tunnel in Northwest Himalaya, India. In Oliveira et al. (eds), *Proc. 7th IAEG congress, Lisbon.* 4347-4354. Rotterdam: Balkema

Dollinger, G., Finnsson, S. & Krauter, D. 1993. An update on the performance of large diameter (483 mm) cutters and high performance TBMs. In Bowerman & Monsees (eds), *Proc. RETC. Boston, MA.* 781-792. Littleton, CO: Soc. for Mining, Metallurgy, and Exploration, Inc.

Einstein, H.H. & Bobet, A. 1997. Mechanized tunnelling in squeezing rock – From basic thoughts to continuous tunnelling. In Golser, Hinkel & Schubert (eds), *Tunnels for People.* 619-632. Rotterdam: Balkema.

Fawcett, D.F. 1993. The effects of rock properties on the economics of full face TBMs. In Hudson et al. (eds), *Comprehensive rock engineering.* Ch. 10(4): 293-311. UK: Pergamon.

Francis, T.E. 1991. Determination of the influence of joint orientation on rock mass classification for tunnelling using a stereographic overlay. *Q. Jl Engng Geol.* 24: 267-273. Northern Ireland: The Geological Society.

Garshol, K. 1980. Fullprofilboring i ustabilt fjell. [Norwegian. Full face boring in unstable rock masses.] Fjellsprengningsteknikk. Bergmekanikk/Geoteknikk. 4.1-4.15. Trondheim: Tapir Press.

Garshol, K. 1983. Excavation, support and pre-grouting of TBM-driven sewer tunnel. Norwegian Tunnelling Technology. *Norwegian Soil and Rock Engineering Association Publication* 2: 21-28. Trondheim: Tapir Press.

Gehring, K. & Kogler, P. 1997. Mechanized tunnelling: Where it stands and where it has to proceed from a manufacturers view point. In Golser, Hinkel & Schubert (eds), *Tunnels for People.* 651-664. Rotterdam: Balkema.

Gildner, J.P., Nowak, D., Painter, D.Z., Revey, G.F., Wilmoth, P. & Yanagisawa, S. 1997. Punk rock in Portland. In Carlson & Budd (eds), *Proc. RETC. Las Vegas, NA.* 151-182. Littleton, CO: Soc. for Mining, Metallurgy, and Exploration, Inc.

Gowring, I.M. & Dickson, K.R. 1987. Underground work at Calaveras. In Jacobs & Hendricks (eds), *Proc. RETC. New Orleans, LA.* 2:711-734 Littleton, CO: Soc. for Mining, Metallurgy, and Exploration, Inc.

Grandori, R., Jaeger, M., Antonini, F. & Vigl, L. 1995. Evinos-Mornos Tunnel – Greece. Construction of a 30 km long hydraulic tunnel in less than three years under the most adverse geological conditions. In Williamson & Gowring (eds), *Proc. RETC. San Francisco, CA.* 747-767. Littleton, CO: Soc. for Mining, Metallurgy, and Exploration, Inc.

Grandori, R., Sem, M., Lembo-Fazio, A. & Ribacchi, R. 1995. Tunnelling by double shield TBM in the Hong Kong granite. In Fuji (ed.), *Proc. 8th ISRM congress, Tokyo.* 2: 569-574. Rotterdam: Balkema.

Green, M.G. & Wallace, W.A. 1993. Large diameter tunneling in a soft clay shale: A case history of the San Antonio flood control tunnels. In Haimson (ed.), *34th US symposium on rock mechanics, Madison WI.* I: 165-168. [Pre-prints] Univ. of Wisconsin.

Grimstad, E. 1999. Private communication.

Grimstad, E. & Barton, N. 1993. Updating of the *Q*-System for NMT. In Kompen, Opsahl & Berg (eds), *Proc. of the International Symposium on Sprayed Concrete – Modern Use of Wet Mix Sprayed Concrete for Underground Support, Fagernes.* Oslo: Norwegian Concrete Association.

Hansen, A.M. 1998. The history of TBM tunnelling in Norway. *Norwegian Soil and Rock Engineering Association Publication* 11: 11-19. Trondheim: Tapir Press.

Hansteen, H. 1991. NGI project report.

Hendricks, R.S. 1969. Hecla Mining Company raise boring and tunnel boring machine experience. *Proc. RETC. Sacramento, CA.* Littleton, CO: Soc. for Mining, Metallurgy, and Exploration, Inc.

Hoek, E. & Brown E.T. 1980. *Underground Excavations in Rock.* 527 p. London: The Institute of Mining and Metallurgy.

Hunter, P.W. & Aust, M.I.E. 1987. Excavation of a major tunnel by double shielded TBM through mixed ground basalt and clayey soils. In Jacobs & Hendricks (eds), *Proc. RETC. New Orleans LA.* 1: 526-561. Littleton, CO: Soc. for Mining, Metallurgy, and Exploration, Inc.

Innaurato, N., Mancini, R., Rodena, E. & Sampaolo, A. 1987. Tunnel boring with TBM and blasting of a hydroelectric tunnel in South Italy. *Proc. Underground Hydropower Plants, Oslo.* 1001-1013. Trondheim: Tapir Press.

Innaurato, N., Mancini, R., Rodena, E. & Zaninetti, A. 1991. Forecasting and effective TBM performances in a rapid excavation of a tunnel in Italy. In Wittke (ed.), *Proc. 7th ISRM congress, Aachen.* 2: 1009-1014. Rotterdam: Balkema.

Innaurato, N., Mancini, R., Rodena, E. & Zaninetti, A. 1994. Comparison of two classification systems as applied to the Alpe Devero tunnel, Italy. *Proc. 7th International Symposium: Tunnelling '94.* 39-49. Chapman & Hall.

ISRM 1981. Suggested methods for rock characterization, testing and monitoring. Brown (ed.). Oxford: Pergamon Press.

Janzon, H. & Buechi, E. 1987. Record performance of TBM in the 13.5 km Amlach Tunnel, Austria. In Jacobs & Hendricks (eds), *Proc. RETC. New Orleans LA.* 2: 1251-1268. Littleton, CO: Soc. for Mining, Metallurgy, and Exploration, Inc.

Japanese Highways 1994. Brochure for Tomei II Highway. NGI 1994. Contract report.

Johannessen, S. 1998. The Meråker Project – 10 km of tunnel in 12 months. *Norwegian Soil and Rock Engineering Association Publication* 11: 85-89. Trondheim: Tapir Press.

Johannessen, S. & Askilsrud, O.G. 1993. Meraaker Hydro – Tunnelling the 'Norwegian Way'. In Bowerman & Monsees (eds), *Proc. RETC. Boston, MA.* 415-429. Littleton, CO: Soc. for Mining, Metallurgy, and Exploration, Inc.

Kadkade, D.G. 1998. Rock excavation – underground structures. *Proc. Int. Conf. on Hydro Power Development in Himalayas, Shimla, India.* 165-177. New Dehli: Oxford and IBH Publishing Co. PVT. Ltd.

Kaynia, A. 1995. NGI contract report.

Klein, S., Schmoll, M. & Avrey, T. 1995. TBM performance at four hard rock tunnels in California. In Williamson & Gowring (eds), *Proc. RETC. San Francisco, CA.* 61-75. Littleton, CO: Soc. for Mining, Metallurgy, and Exploration, Inc.

Korbin, G.E. 1998. Claims and tunnel boring machines: Contributing factors and lessons learned. In Moore & Hungr (eds), *Proc. 8th IAEG congress, Vancouver.* 3523-3528. Rotterdam: Balkema.

Kovári, K. 1998. Tunnelling in Switzerland. *Tribune No. 5. ITA-AITES – January 1998.* 1-23.

Løset, F. 1992. Support needs compared at the Svartisen road tunnel. *Tunnels & Tunnelling, June 1992.* UK: British Tunnelling Society.

Løset, F., Barton, N., Grimstad, E. & Kveldsvik, V. 1996. The Q-method used in TBM tunnels, field mapping and core logging. *Rock Mechanics Conference, Stockholm.* pp 65-78. Svensk Bergteknisk Forskning.

Martin, D. 1988. TBM tunnelling in poor and very poor rock conditions. *Tunnels & Tunnelling, March, 1988.* 22-28. UK: British Tunnelling Society.

Martinsen, J. 1987. Bergen and beyond: Western Norway's plans for a tunnel and bridge network. In Jacobs & Hendricks (eds), *New Orleans, LA.* 2: 1220-1233. Littleton, CO: Soc. for Mining, Metallurgy, and Exploration, Inc.

McCormick, B., Gilmore, E., Tudor, R., Bush, B. & Korbin, G.E. 1997. Analysis of TBM performance at the record setting River Mountains tunnel #2. In Carlson & Budd (eds), *Proc. RETC. Las Vegas, NA.* 135-149. Littleton, CO: Soc. for Mining, Metallurgy, and Exploration, Inc.

McFeat-Smith, I. 1981, Machine tunnelling in weak and fractured rock conditions. *Proc. Int. symp. on weak rock, Tokyo.* 1075-1080.

McFeat-Smith, I. & Askilsrud, O.G. 1993. Tunnel boring machines in Hong Kong. In Bowerman & Monsees (eds), *Proc. RETC. Boston, MA.* 401-413. Littleton, CO: Soc. for Mining, Metallurgy, and Exploration, Inc.

McKelvey, J.G., Schultz, E.A., Helin, T.A.B. & Blindheim, O.T. 1996. Geotechnical analysis in S. Africa. *World Tunnelling, November 1996.* 377-390.

Milanovic, P. 1997. Tunnelling in karst: Common engineering-geology problems. In Marinos, Koukis, Tsiambaos & Stournaras (eds), *Engineering Geology and the Environment.* 2797-2802. Rotterdam: Balkema.

Mitani, S. 1998. The state of art of TBM excavation and probing ahead technique. In Moore & Hungr (eds), *Proc. 8th IAEG congress, Vancouver.* 3501-3512. Rotterdam: Balkema

Mitani, S., Iwai, T. & Isahai, H. 1987. Relations between conditions of rock mass and TBM's feasibility. *Proc. 6th ISRM congress, Montreal.* 1: 701-704. Rotterdam: Balkema.

Morimoto, T. & Hori, M. 1986. 6. Performance characteristics of a tunnel boring machine from the geomechanical viewpoint. *Int. J. Rock Mech. Min. Sci. & Geomech. Abstr.* 23(1): 55-66. UK: Pergamon.

Movinkel, T. & Johannessen, O. 1986. Geological parameters for hard rock tunnel boring. *Tunnels & Tunnelling, April, 1986.* 45-48. UK: British Tunnelling Society.

Nelson, P.P. 1993. TBM performance analysis with reference to rock properties. In Hudson et al. (eds), *Comprehensive rock engineering.* Ch. 10(4): 261-291. UK: Pergamon.

Nelson, P.P. 1987. Soft rock tunnelling: Equipment selection concepts and performance case histories. In Jacobs & Hendricks (eds), *New Orleans LA. New Orleans, LA.* 1: 583-600. Littleton, CO: Soc. for Mining, Metallurgy, and Exploration, Inc.

Nelson, P.P. 1996. Rock engineering for underground civil construction. In Aubertin, Hassani & Mitri (eds), *2nd NARMS '96. Montréal, Québec.* 1: 3-12. Rotterdam: Balkema.

Nelson, P., O'Rourke, T.D. & Kulhawy, F.H. 1983. Factors affecting TBM penetration rates in sedimentary rocks. In Mathewson (ed.), *24th US symposium on rock mechanics, Texas A&M Univ.* 227-237. College Station, TX: Association of Engineering Geologists.

Nelson, P.P., Al-Jalil, Y.A. & Laughton, C. 1999. Improved strategies for TBM performance prediction and project management. In Hilton & Samuelson (eds), *Proc. RETC. Orlando, FL.* Ch. 54: 963-979. Littleton, CO: Soc. for Mining, Metallurgy, and Exploration, Inc.

NGI 1991. Contract report.

NGI 1997. Contract report.

NGI 1998. Contract report.

NGI 1999. Contract report.

Nilsen, B. & Ozdemir, L. 1993. Hard rock tunnel boring prediction and field performance. In Bowerman & Monsees (eds), *Proc. RETC. Boston, MA.* 833-852. Littleton, CO: Soc. for Mining, Metallurgy, and Exploration, Inc.

Nishioka, K. & Aoki, K. 1998. Rapid tunnel excavation by hard rock TBMs in urban areas. In Negro Jr & Ferreira (eds), *Tunnels and Metropolises.* 655-661. Rotterdam: Balkema.

Nordmark, A.M. & Franzen, T. 1993. Subsurface space – an important dimension in Swedish construction. In Hudson et al. (eds), *Comprehensive rock engineering.* Ch. 2(5): 29-54. UK: Pergamon.

NTH 1994. Full face tunnel boring. [In Norwegian: Fullprofilboring av tunneler.] Prosjektrapport anleggsdrift 1-94. Trondheim: NTH.

Palmström, A. 1982. The volumetric joint count – a useful and simple measure of the degree of rock mass jointing. *Proc. 4th IAEG Congress, New Delhi.* V: Theme 2: 221-228.

Pikering, R.G.B., Watson, I.C., Klokow, J.W. & Knoetze, A.F. 1999. Practical feasibility of using TBMs in deep level gold mines. In Hilton & Samuelson (eds), *Proc. RETC. Orlando, FL.* Ch. 55: 981-992. Littleton, CO: Soc. for Mining, Metallurgy, and Exploration, Inc.

Rautenberg, R. & McDermott, P. 1997. TARP – The Des Plaines tunnel system (North leg) A case history. In Carlson and Budd (eds), *Proc. RETC. Las Vegas, NA.* 183-200. Littleton, CO: Soc. for Mining, Metallurgy, and Exploration, Inc.

Rienößl, K. 1987. Mechanical tunnelling method applied to gneiss. *Proc. Underground Hydropower Plants, Oslo.* 953-964. Trondheim: Tapir Press.

Robbins, R.J. 1982. The application of tunnel boring machines to bad rock conditions. *ISRM symposium, Aachen.* 2: 827-836. Rotterdam: Balkema.

Robbins, R.J. 1997. Hard rock tunneling machines for squeezing rock conditions: Three machine concepts. In Golser, Hinkel & Schubert (eds), *Proc. ITA Congress, Vienna. Tunnels for People.* 633-638. Rotterdam: Balkema.

Rostami, J. & Ozdemir, L. 1993. A new model for performance prediction of hard rock TBMs. In Bowerman & Monsees (eds), *Proc. RETC. Boston, MA.* 793-809. Littleton, CO: Soc. for Mining, Metallurgy, and Exploration, Inc.

Sandtner, A.K. 1993. Clermont Tunnel, Switching a problem-TBM-site into a success. In Bowerman & Monsees (eds), *Proc. RETC. Boston, MA.* 889-904. Littleton, CO: Soc. for Mining, Metallurgy, and Exploration, Inc.

Sanio, H.P. 1985. Prediction of the performance of disc cutters in anisotropic rock. *Int. J. Rock Mech. Min. Sci. & Geomech. Abstr.* 22(3): 153-161. UK: Pergamon.

Scesi, L. & Papini, M. 1997. From pilot tunnel to main tunnel: a study of scale effect. *Gallerie e grandi opere sotterranee.* Luglio 1997. 52: 39-51.

Scolari, F. 1995. Open-face borers in Italian Alps. *World Tunnelling*, November 1995. 361-366.

Sharp, J.C., Warren, C.D., Barton N.R. & Muir Wood, R. 1996. Fundamental evaluations of the chalk marl for the prediction of UK marine tunnel stability and water inflows. In Harris, Hart, Varley, & Warren (eds), *Engineering Geology of the Channel Tunnel.* 472-507. London: Thomas Telford.

Shen, C.P., Tsai, H.C., Hsieh, Y.S. & Chu, B. 1999. The methodology through adverse geology ahead of Pinglin large TBM. In Hilton & Samuelson (eds), *Proc. RETC. Orlando, FL.* Ch. 8: 117-137. Littleton, CO: Soc. for Mining, Metallurgy, and Exploration, Inc.

Singh, B. 1993. Norwegian Method of Tunnelling Workshop, CSMRS, New Dehli.

Skjeggedal, T. 1998. Six case histories. *Norwegian Soil and Rock Engineering Association Publication* 11: 79-83. Trondheim: Tapir Press.

Stevenson, G.W. 1999. Empirical estimates of TBM performance in hard rock. In Hilton & Samuelson (eds), *Proc. RETC. Orlando, FL.* Ch. 56: 993-1009. Littleton, CO: Soc. for Mining, Metallurgy, and Exploration, Inc.

Storjordet, A. 1981. VEAS tunnel system – construction considerations. [In Norwegian.] Fjellsprengningsteknikk. Bergmekanikk/Geoteknikk. 32.1-32.16. Trondheim: Tapir Press.

Sundaram, N.M. & Rafek, A.G. 1998. The influence of rock mass properties in the assessment of TBM performance. In Moore & Hungr (eds), *Proc. 8th IAEG congress, Vancouver.* 3553-3559. Rotterdam: Balkema.

Tarkoy, P.J. & Marconi, M. 1991. Difficult rock comminution and associated geological conditions. *Proc. 6th Int. Symposium: Tunnelling '91, London.* 195-207. Elsevier.

Thuro, K. 1997. Prediction of drillability in hard rock tunnelling by drilling and blasting. In Golser, Hinkel & Schubert (eds), *Proc. ITA Congress, Vienna.Tunnels for People.* 103-108. Rotterdam: Balkema.

Toolanen, B., Hartwig, S. & Janzon, H. 1993. Design considerations for large hard rock TBMs when used in bad ground. In Bowerman & Monsees (eds), *Proc. RETC. Boston, MA.* 853-868. Littleton, CO: Soc. for Mining, Metallurgy, and Exploration, Inc.

Tseng, Y.Y., Tsai, H.C., Tseng, C.T. & Chu, B. 1998. The Pinglin eastbound mechanized tunnelling. In Negro Jr & Ferreira (eds), *Proc. ITA Congress, São Paulo: Tunnels and Metropolises.* 787-792. Rotterdam: Balkema.

Tseng, Y.Y., Wong, S.L., Chu, B. & Wong, C.H. 1998. The Pinglin mechanised tunnelling in difficult ground. In Moore & Hungr (eds), *Proc. 8th IAEG congress, Vancouver.* 3529-3536. Rotterdam: Balkema.

VanDerPas, E. & Allum, R. 1995. TBM technology in a deep underground copper mine. In Williamson & Gowring (eds), *Proc. RETC. San Francisco, CA.* 129-143. Littleton, CO: Soc. for Mining, Metallurgy, and Exploration, Inc.

Varley & Warren, C.D. 1996. Chapter 2: History of the geological investigations for the Channel Tunnel. In Harris, Hart, Varley & Warren (eds), *Engineering Geology of the Channel Tunnel.* 5-18. London: Thomas Telford.

Voirin, J. & Warren, C.D. 1996. Chapter 15: French tunnels: geotechnical monitoring and encountered conditions. In Harris, Hart, Varley & Warren (eds), *Engineering Geology of the Channel Tunnel.* 244-260. London: Thomas Telford.

Wallis, S. 1998. Pinglin perseverance in Taiwan. *Tunnel 7/98.* 10-24. Köln: STUVA.

Wanner, H. & Aeberli, U. 1979. Tunnelling machine performance in jointed rock. *Proc. 4th ISRM congress, Montreux.* 1: 573-579. Rotterdam: Balkema.

Wanner, H. 1980. Stability problems by fullface tunnel boring. Fjellsprengningsteknikk. Bergmekanikk/Geoteknikk. 25.1-25.7. Trondheim: Tapir Press.

Warren, C.D., Varley, P.M. & Parkin, R. 1996. Chapter 14: UK tunnels: Geotechnical monitoring and encountered conditions. In Harris, Hart, Varley, & Warren (eds), *Engineering Geology of the Channel Tunnel.* 219-243. London: Thomas Telford.

Watanabe, K., Sunamichi, M. & Kajiyama, T. 1991. Tunnel excavation with TBM pilot machine on AKIHA No. 3 hydropower station. In Wittke (ed.), *Proc. 7th ISRM congress, Aachen.* 2: 1023-1027. Rotterdam: Balkema.

Wickham, G.E., Tiedemann, H.R. & Skinner, E.H. 1974. Ground support prediction model. RSR concept. NTIS Report, San Francisco.

World Tunnelling 1991. High production at Svartisen. WT, Nov. 1991, 339-344.

Zell, S. 1995. Final stage of Storebælt railway tunnel – Summary and appraisal. In Williamson & Gowring (eds), *Proc. RETC. San Francisco, CA.* 775-777. Littleton, CO: Soc. for Mining, Metallurgy, and Exploration, Inc.

Index

9 789058 093417

Printed and bound by CPI Group (UK) Ltd, Croydon, CR0 4YY

23/10/2024

01777679-0011